ENGAGING CUSTOMERS USING BIG DATA

ENGAGING CUSTOMERS USING BIG DATA

HOW MARKETING ANALYTICS ARE

TRANSFORMING BUSINESS

ARVIND SATHI

palgrave
macmillan

First published in 2014 by
PALGRAVE MACMILLAN®
in the United States—a division of St. Martin's Press LLC,
175 Fifth Avenue, New York, NY 10010.

Where this book is distributed in the UK, Europe and the rest of the world,
this is by Palgrave Macmillan, a division of Macmillan Publishers Limited,
registered in England, company number 785998, of Houndmills,
Basingstoke, Hampshire RG21 6XS.

Palgrave Macmillan is the global academic imprint of the above companies
and has companies and representatives throughout the world.

Palgrave® and Macmillan® are registered trademarks in the United States,
the United Kingdom, Europe and other countries.

ISBN: 978–1–137–38618–2

Library of Congress Cataloging-in-Publication Data

Sathi, Arvind.
    Engaging customers using big data : how marketing analytics are
    transforming business / Arvind Sathi.
        pages cm
        ISBN 978–1–137–38618–2 (hardback)
        1. Marketing—Management. 2. Marketing—Data processing.
    3. Internet marketing. 4. Consumer behavior. 5. Marketing research.
    6. Big data. I. Title.
HF5415.13.S273 2014
658.8'3—dc23   2014000959

A catalogue record of the book is available from the British Library.

Design by Newgen Knowledge Works (P) Ltd., Chennai, India.

First edition: July 2014

10 9 8 7 6 5 4 3 2

Printed in the United States of America.

To my social media friends who liked, poked, friended, and shared stories with me all these years.

CONTENTS

ILLUSTRATIONS

FIGURES

TABLE

ACKNOWLEDGMENTS

First and foremost, I would like to acknowledge the hard work on the CMO study conducted by IBM, as well as the related case studies and related solutions carried out by the IBM Software Group Big Data marketing teams. I have used publicly referenced material, including the use cases, the Solution Architecture framework, and a number of case studies, and have provided additional references for the audience to explore further. IBM senior management has paved the way with an inspiring vision of analytics-driven CMO of the future, especially Ginni Rometty's talk at the CMO+CIO conference. Jeff Jonas provided me with inspiration for experimenting with the ideas and continues to be my source of inspiration. The algorithms and architectures for marketing analytics were created with help from Barry Rosen, Bob Johnston, Daryl BC Peh, Eberhard Hechler, Elizabeth Dial, Hamid Pirahesh, Heng Cao, Ken Babb, Mandy Chessel, Manav Gupta, Nagui Halim, Noman Mohammed, Paul Christensen, Paul Colgan, Peter Harrison, Rich Harken, Sambit Sahu, Sandy Tucker, Tapasi Sengupta, Tommy Eunice, and Yoel Arditi. The Dallas Global Solutions Center team—Christian Loza, Mathews Thomas, and Janki Vora—provided valuable experimentation on the ideas. Anthony Behan, Audrey Laird, Bob Weiss, Christine Twiford, Craig Ginsberg, Dirk Michelsen, Emeline Tjan, Ken Kralick, Laura Lederer, Livio Ventura, Mehul Shah, Neil Isford, Norbert Herman, Oliver Birch, Perry McDonald, Raquel Katigbak, Richard Lanahan, Richard Maraschi, Rick Flamand, Robert Segat, Sara Philpott,

Steve Cohen, Steve Teitzel, Steve Trigg, Tushar Mehta, and Wolfgang Bosch helped identify opportunities and business value.

Next, I would like to acknowledge Adam Gersting, Anu Jain, Aparna Betigeri, Carmen Allen, Darin Briskman, Jinesh Radadia, John Held, Joe Milich, Joseph Baird, Judith List, Kedrick Brown, and Piyush Sarwal, who helped me understand implementation issues and organizational implications. I would like to thank Roger Rea, Dan Debrunner, Vibhore Kumar, Joey Foley, Shankar Venkataraman, Girish Venkatachaliah, Karthik Hariharan, Claudio Zancani, and Mike Zucker for providing me with insights into commercially available products and how they can be used for big data analytics.

I was fortunate to interact with a number of practitioners as I studied big data opportunities in marketing. This includes Ash Kanagat, Dave Dunmire, Doug Dickerson, Gautam Shah, Girish Varma, Harpinder Singh Madan, Harsch Bhatnagar, Jamie Williams, Jamil Husain, Jennifer Lim, Jessica Ma, Joshua Koran, Kanishka Agarwal, Maureen Little, Mike Cooley, Philip Smolin, Ravi Kothari, Rob Smith, Sumit Chowdhury, Sumit Singh, Teresa Jacobs, Una Fox, and Von McConnell. I am grateful for the insightful discussions and implementations in understanding business opportunities as well as current big data practices.

I would like to thank the Palgrave Macmillan editorial team for their review of the book. Gaurav Deshpande did a fair amount of work with me in creating the cartoons used here. I would like to thank Slice and Turn marketing organizations for helping me with the illustrations used in this book.

I received a fair amount of feedback from my book titled *Big Data Analytics* that was published earlier. I would like to thank the audience for their valuable feedback, which helped me shape this book. I hope the feedback makes this book more readable and improves its usability.

Last, but not least, I would like to thank my wife Neena, my daughter, Kinji, my son-in-law, Kevin, and my son, Conal, for their inspiration, support, and editorial help.

1

INTRODUCTION

WHY THIS BOOK?

As I was watching a movie online just before US presidential election last year, the website displayed an advertisement every ten minutes. Since I had not paid anyone for watching the movie and am used to watching commercials on television, it should not have been unusual to see a commercial every ten minutes or so. The website, however, showed me the same commercial over and over. After about the eighth time, I felt sorry for the poor advertiser (someone advertising for Mitt Romney as presidential candidate) because the effectiveness of the ad had long since dissipated and, instead, an annoyance factor had crept in. I was facing a real-time decision engine that was rigid and was placing an advertisement without any count or analysis of saturation factor. As an aside, I lived in a "swing state" for the fall 2012 US presidential elections, so it is possible the advertising agency for the candidate had decided to saturate the advertisements at my location. I must confess, I am not very politically aligned with either party. However, I do have a curiosity about marketing analytics and how political candidates market themselves. So, I decided to respond to the Obama campaign to compare the two. I was amazed to see a level of personalization in the campaign. Unlike the push campaign from Romney, the Obama campaign team worked hard to collect and document my personal political preferences, and knew how to target his responses personally to me in emails. His 2012 campaign is a grand example of marketing analytics driven by big data, and was studied in detail, even by the Republican Party. The campaign used an interesting mix of big data analytics with many connection points with the voters. The campaign drew heavily on big data analytics to drive email campaigns, television spot purchases, and even in scheduling President Clinton's visits to battleground counties.[1] To compare, I have since subscribed to a couple of other political campaigns, including one for the Tea Party. In most cases, the campaigns are using emails to push what they always did with junk mail, that is, a set of standard messages.

So much for political marketing, how are the consumer marketers dealing with big data driven collaboration with their customers? Figure 1.1 shows a caricature of how someone may initiate their social medial interaction. How would a marketer sense customer need for information and respond accordingly? Let me go back to the advertising on the website, where I was watching the movie. How could the website offer advertisements during breaks, which I would find relevant? For example, the site was well aware of the movie genre I had accessed during several visits to the site. An analysis of this genre could have placed me in several viewing segments. In fact, the same website offers me movie recommendations, which are based on prior viewing habits. The viewing segments could be of tremendous value to the advertisers in advertising decisions. After watching the movie for a while and repeatedly viewing the same advertisement, I decided to take a break from watching the movie to searching for a food processor as a gift to my son. When I returned to the movie, I again faced the same advertisement. I had secretly hoped that my food processor search would conveniently trigger an advertisement for a good food processor to help me in my purchase. By sensing and analyzing my previous web searches, marketers could have offered me appropriate information or promotions, thereby increasing the advertisement relevance for me. This book will show you how a marketer can sense a customer's needs and respond in real time

Figure 1.1 How was your first marketing exposure to the Social Media?

with a campaign that focuses on customer needs and offers a customized campaign for a solution that will meet the need.

As an evangelist for big data analytics, I have been in front of audiences with different levels of maturity and excitement about big data availability and use. These audiences have raised a number of good questions about big data and its impact on marketing, which inspired me to document the brainstorming and ideas discussed in response to those questions. This book covers a set of fundamental questions you may have at the back of your minds. As I have watched the changes that have taken place over the last 35 years, the questions below have reflected my own curiosity about big data. Certainly, the field is rapidly evolving, and I have done my best to highlight the maturity in our collective understanding of the field, which impacts the certainty of our responses.

What has changed marketers' communication with their prospects and customers? In the early 1980s, in my first year of immigration to the United States, I found in my mailbox a big envelope from American Family Publishers (AFP) with my name printed in a 72-point font, "Arvind Sathi, you have won a ten million dollar a jackpot, you only need to subscribe to the magazines listed here." I was intrigued that the marketing systems had caught up with me so fast and that they could spell my name correctly, find my address, and offer a chance to win money. However, despite AFP's attempt at personalization, other people around me received identical packages, and it was ironic that AFP offered the same magazines to everyone! While AFP learned to spell my name and research my address, they did not have a customized set of magazines for my needs. For the big $10,000,000 sweepstakes in 1985, New York state employee Lillian Countryman calculated the odds of winning. Players of the AFP sweepstakes had a 1 in 200,000,000 chance.[2] As B-school graduate students, a group of us had a healthy debate regarding how to make such offers appealing to consumers, and we concluded that customization of a magazine list was a better way to attract customers, as opposed to offers for an elusive jackpot. We also realized that the cost of customization at that time was prohibitive, as AFP had no easy way of collecting micro-segmentation information about their target customers. Although the modus operandi has not changed for a number of marketers, there are subtle changes in the wind. Consumers continue to receive a large collection of junk postal mails and emails, but they now have filters for most of it. Based on the breadcrumbs available from consumers, savvy marketers are increasingly fine-tuning their targeting. I am currently in the process of buying a new house. I am amazed to find a new set of catalogs offering me outdoor furniture. While I discard most of the junk mail, I find myself studying these catalogs, marveling at the marketing process at Frontgate for appropriately targeting me as a recipient for their catalog. Many catalog marketers are able to establish a dialogue with their customers, to offer additional information related to items in a catalog, where the customer has shown interest. How do these marketers sense and respond to specific customer needs, and what type of attention are they able to attract from their customers? What are the analytics capabilities that enable these marketers to be so focused and conversational with their customers?

How is marketing research changing with big data and related technological forces? Marketers have been among of the most sophisticated users of social research and have invested a fair amount into gathering data from their customers. In the past, the limiting factor was the number of observations collected for marketing research. National surveys were hard to collect and required a massive investment of time and resources. In addition, surveys collected past recollection of customer choices as reported by the customers. The data was only as good as the sampling technique and size. It relied heavily on the survey administrator's ability to ask questions, and respondent's ability to recollect history. Now the floodgates are gradually opening, as what was formerly the wastebasket in corporate information technology (IT) departments and consumer personal computers (PCs) is becoming a gold mine for marketers and market data traders. Big data analysts are lining up with buckets, gathering all the bits they can find anywhere to collect and analyze past events. What do these bit buckets offer to marketers?

How are these new sources of data and advanced techniques for analytics changing the way we conduct research and analyze research data?

What do I do to the decades of investments in marketing partnerships, processes, skills, and technologies? Is it revolutionary or evolutionary change? Marketing received its due share of resources and investments over several decades. In addition, a number of external organizations helped marketers buy the data and the instruments for communications—for example, the prime-time television viewership data. Marketing processes were built to take advantage of the available resources, with a razor-sharp focus on optimizing a set of measurable key performance indicators (KPIs). In any large Fortune 50 organization, there are hundreds of skilled resources, as well as a large number of computing resources dedicated to marketing analytics or sources of data, which feed marketers with the required source information. Over the last decade, we have witnessed the development of a new marketplace. On the one hand, it is changing our current processes and organizations. On the other hand, it is challenging some of the fundamental principles and removing many hard constraints. Instead of buying advertisement space in advance, much of the online advertising spots are being claimed through auction platforms. A new breed of analysts has emerged with a new set of technologies for big data analytics. With a myriad of cloud-based data sources, and third parties offering social media interactions with customers, how do we change our partnerships, processes, skill sets, technology mix, and our decade of investment in marketing science tools and techniques? How do these factors change marketing research, advertising, pricing, and product management organizations? It seems like each evolution of technology seeks a replacement for everything we have achieved to date. How would a marketer continue to evolve these processes as new marketing data vendors show up with clouds and big data?

Is big data a big hype created by a handful of social media companies that will fizzle once customers and marketers understand the privacy implications? What is the longevity of this wave and is there a crash coming? In every presentation I have given, I find a couple of skeptics, someone who is ready to challenge the big data tsunami. How about the new sources of data? Is that not adding a lot of lies from which is hard to extract any additional truth? And how about customer trust? Are we likely to lose our customer base as we eavesdrop on their behavior? What happens if consumers get tired of it and stop using the engines that drive big data? What the skeptics address mostly are the well-known, and well-publicized areas of concern. The market is still maturing and a fair number of processes have yet to be discovered and properly regulated. When the marketers discovered short messaging service (SMS) on the mobile phone, there was a plethora of advertising initiated using SMS. However, the Telecom Consumer Protection Act in the United States and similar laws in many Asian countries have issued new guidelines that impose restrictions on telemarketers using SMS for messaging[3] and require explicit opt-in.[4] Marketers must conform to these regulations or pay hefty fines. The guidelines give us a glimpse of how regulations will help consumers use technology without runaway misuse. In addition, there is a much greater number of bit buckets, which can be used without sparking any controversial issues. Also, there are ways to use the data, which is fairly legitimate and respectful of consumer privacy concerns. The market has yet to fully mature, as marketers evolve ways to effectively interact with customers using new technologies.

I do not want to understand big data. Can someone explain what it does to marketing? Almost every other book I have read gets to the techniques of big data too fast, without delving into marketing analytics. While there are significant technological advancements, the real changes are sociological and organizational. They are reflected in the hefty market valuations for the new information providers. The organizational relationships are rapidly changing. Consumers have figured out how to fast-forward through the push marketing and avoid any messaging they do not want. Telecom organizations are finding they are at the epicenter of shopping and content viewing, and are very interested in the monetization of their data to retailers. Auto manufacturers are rapidly designing connected cars, which should be aptly renamed "mobile devices." Insurers would like to offer insurance based on what we eat and how we drive our cars. I am grateful to my editor for pushing me hard to write this book for the marketers and not for the technologists. The lenses this book has applied are those of marketers. The focus is on how changes to the consumers and the environment are shifting the basic function of marketing, and how the analytics will reshape how we market to these consumers.

How about business-to-business (B2B) marketing? Most discussions are about consumer markets. Is there anything changing about B2B marketing? How can big data be applied to B2B marketing? Also, how is big data shared across industries? Corporate marketing is going through its own silent revolution. Marketers and customers are changing how they interact with each other. Social media is radically altering how professionals interact. Industrial research has many more avenues for data gathering. YouTube is rapidly emerging as the platform for corporate messaging and product demonstrations. Blogs are increasingly being used as ways to research ideas or to initiate and promote product buzz. A fair amount of IT marketing is directed to corporate customers. How do corporate marketers keep track of their customers, their needs, competitive activities, and major changes?

PROPOSITIONS

Big data is rapidly transforming a number of business functions across many industries. The biggest shift is in how we market to our customers. Unlike yesterday's environment, where marketers broadcasted across a set of customer segments, we can now personalize the communications to each customer based on his/her current predisposition to the products being sold. I have been observing the growth of big data and advanced analytics and its impact on marketing analytics. I would like to make three propositions, which summarize my understanding of how big data and advanced analytics will change marketing analytics. These propositions were inspired by a talk IBM CEO Ginni Rometty gave to CMO community[5] and have evolved as I researched the related topics and discussed with a number of practitioners.

Proposition 1: From "Sample recalls" to "Observing the Population": For many years, marketers sought small samples from their customers, used interviews to recall past behaviors, and extrapolated to reconstruct the buying behavior for the larger population. Marketers learned to create a number of sophisticated techniques for projecting results to the larger population using the smallest samples they could afford, as it was cost prohibitive to collect and store large data. Also, marketers had to conduct interviews in limited amount of time, leading to broad questions and inaccurate results. Marketers did not have the luxury of asking detailed questions, like how many products did you browse before making the purchase selection, and whether they were all similar or very different. Big data has changed the game completely. We can connect with customers, record every click on the Web, watch every step in the store, and listen to all the public conversations. Big data brings many more observations to the table, thereby leading to a lot more detailed analysis and discovery. Marketers now have access to a excruciatingly detailed understanding of shopping behavior. In addition, marketers are no longer dealing with small samples as we now have access to customer data for a very large proportion of the overall population. How does it change our understanding of the consumer? We need an improved vocabulary and a new way of interpreting statistical data. Let me use a sports analogy to illustrate this proposition. When I was about 12 years old, my father first showed up to watch our cricket game. While he was trying to understand the game, he became far more fascinated with the scorekeeping process. "I wish they would do this level of detailed scorekeeping at work, and we would have a lot more objective performance evaluation," he told me as he watched us painstakingly record every ball and score. Somehow, most sports, such as American football, soccer, tennis, or cricket, carry their own versions of very well-defined methods for creating a great number of detailed observations. Imagine an analysis of American football without passing yards, touchdown passes, and interceptions. Without this statistic, we could still have the resulting scores, but would not be able to deduce the quality of the game. A sample recall is like knowing the final score, while a detailed observation is like knowing the passing yards and the interception information that contributed to the final scores. Imagine a set of researchers standing outside stadiums asking the audience leaving a game how they would differentiate the winning team in the absence of these observed statistics. The observations have provided a vocabulary and a way of comparing players and teams, and this information has been available long before automation brought us closer to big data in other areas. As automation goes up and storage cost goes down, marketers are seeking an exponential rise in observations in many areas, creating a new set of metrics to define and measure customer needs and interests. Product automation has enabled corporations to collect a lot more data, and corporate IT organizations are getting better at storing rather than deleting all the event-level details.

Proposition 2: Marketing through collaborative Influence: Marketers are seeing a rapid rise in collaborative influence from marketers to customers, from customers to marketers, and from customers to other customers, each of which can be fine-tuned to a large number of microsegments and used for personalized communications. Marketers can converse with the customers as they make decisions, and influence their decision-making using a series of sophisticated marketing tools. Customers can influence how new products are developed and how existing products are evaluated. Marketing in the broadcast era was all about repeating the same message over and over, until it reached desired reach and opportunity to see (OTS). Elias St. Elmo Lewis, cofounder of the Association of National Advertisers and one of the first advocates in the field of advertising, created the AIDA model (awareness, interest, desire, action) around the beginning of the twentieth century, which described how marketers influence customers.[6] In the early days, it was not possible to observe who saw the communication and how they reacted to it. Today's collaborative influence is far more of a two-way persuasion process. Analytics allows marketers to identify customer needs, initiate a dialogue, and customize a collaborative process to converse with customers, whether with the help of experts, targeted advertising, or through personalized communications directly with the customers. Also, crowdsourcing is increasingly used for product idea generation or evaluation. Many product ideas are based on ideas or feedback from the customers. It is possible to listen to customers and act on the communication received.

Proposition 3: From silo'ed to orchestrated marketing: Various marketing departments, such as marketing research, advertising, product management, and sales, had limited avenues for orchestration. We did not know that a specific customer was interested in purchasing a product, and had been to various websites, and hence was a qualified prospect for additional information or promotion. So, the marketing effort could not be targeted and orchestrated across the departments. Using big data and collaborative influence, however, the market leaders are beginning to orchestrate their marketing investment, and focusing their attention on their customers in ways we have never seen before. The four P's—product, pricing, place, and promotion—are no longer offered via a set of silos broadcasting their overlapping messages. A set of sophisticated orchestration engines takes into account customer privacy preferences, needs, intentions, and the current relationship with the marketer to coordinate a set of actions. It is like having an episode of a television sitcom that can be changed based on viewer preferences. The observations are generated using a series of sources, most of them external to the organization. The messaging is delivered via a complex web of marketing organizations, each specializing in a specific instrument. However, the marketer keeps track of customer status and directs the messaging accordingly. If the customer is searching for a product, he/she may at his/her own initiative starting pulling the information. As marketers discover an unmet customer need, they are able to respond back with marketing information, up to and not exceeding a saturation point. If I am shopping for a smartphone, the minute I start searching

the Web, tweeting to my community, or shopping at stores, the marketers can collect breadcrumbs to find my specific needs, preferences, and
constraints, and can tailor messaging to me. As needs change, so does
messaging to keep up with the changes. Once I purchase the smartphone, I should stop receiving those messages. Various touchpoints
keep in touch with each other, so the call center and website are aware
of advertising placed / responded to, and can customize their interactions based on the customer actions already known to them.

DATA SOURCES

I have been a practitioner in marketing analytics for a couple of decades.
In the course of my work, I have conducted consulting assignments,
implemented systems, marketed my products, and studied a number
of market leaders. My current job provides me a great opportunity to
travel worldwide and conduct workshops with a number of companies.
In many situations, I worked with a series of very knowledgeable experts
who took years to perfect many substantive issues, such as customer
privacy. A vast majority of the data presented in this book is based on
these interactions using detailed working sessions from the Americas,
Europe, the Asia-Pacific, the Middle East, Africa, and South America.
My experience has been mostly with telecom providers (landline and
wireless)[7], cable operators, media content owners, information services,
retailers, and utilities companies. I am greatly indebted to the market
leaders, who have shared ideas and experiments. To protect confiden-
tial information, I have only used publicly available material from these
marketers, as well as generic industry practices, which have been prevalent with many market leaders in a specific industry.

The second set of data source is my personal experience as a consumer. I am still a teen at heart and have been a guinea pig for new
experimentation. Facebook with its academic roots only allowed members with an .edu email address in the first couple of years. Luckily,
my alma mater offered an email address, and it gave me the privilege
of joining Facebook in the early days. In this initial period, the only

friends I had on Facebook were my kids, nephews, nieces, and their friends. I guess it gave me a running start over most people, and gave me a chance to start using Facebook as a way to share information with the next generation. In doing so, I tried emulating my two kids and their friends, who are far better than I am in their use of social media. It started as fun sessions with my son's friends, who are all aspiring big data gurus, as we tried making my hula-hooping video viral on YouTube.[8]

The third set of data sources is from my esteemed colleagues at IBM, or business partners, who have been working hard at big data analytics. A number of my examples are based on work others have shared with me. At the time of writing, big data is still in its infancy, and it is fortunate that IBM has given so much emphasis to big data and has invested a fair amount of resources to understand the impact on processes, businesses, and IT technologies. The book carries a fair amount of publicly stated visionary statements from IBM or business partners. It has been my fortunate experience to meet and work with a large number of visionaries who have directed and shaped the technologies underlying marketing analytics.

Last but not least, this is a good time to search for material on marketing analytics. There are a large number of white papers, research findings, and blogs available from a variety of sources. For example, LinkedIn offers many special groups with a vast number of blogs on related topics. By participating in these blogs and reading comments from experts around the globe, I gathered priceless insight. This source is truly crowdsourced and is the largest gift to me from the world of big data.

AUDIENCE

This book is primarily written for marketing professionals. There are a number of professionals who have a great understanding of marketing analytics and how big data is changing their world. I hope this book provides them with a level of comfort that there are many others who are experimenting in similar ways. This book is intended as a conversation-starter with this community. I have done my level best to discuss technological topics in ways marketing people would identify with.

A second audience is the sellers of information technology solutions to the marketers. Many IT sellers, whether IT departments or salespeople working for technology companies, may have a better understanding about the IT components described in this book. However, I hope the book provides them with better insight on the use cases for their tools and solutions. Most IT marketers are doing their best to learn as much as possible about their customers in order to do a better job of connecting to them. This book should provide fuel to this community by offering a synopsis of how marketing analytics is reshaping their customer base, and how marketers are using IT as a lever to change their processes and organizations.

The third audience for the book is the academic professionals who are teaching and learning about marketing science. They have a much better understanding of the principles referred here. The book should provide them with examples that either validate or challenge these principles. I also hope my examples and use cases will provide fertile ground for academic research in marketing.

BOOK OVERVIEW

Using a series of case studies from pioneers, the book will employ the three propositions to show how each marketing function is undergoing fundamental changes: how personalized advertising is delivered using online channels where the marketers identify the specific customer and tailor their messaging based on customer behavior, context, and intention; how customer behavior is collected from a variety of sources across many industries and examined to identify micro-segments; how online and physical stores collaborate to provide a unified shopping experience and deliver product information. Then the book examines the tools and techniques for marketing analytics that support these capabilities. It projects the impact on statistical techniques, qualitative reasoning, and real-time pattern detection, to name a few. Based on these changes, the book prescribes the changes needed to update our skills and tools for marketing analytics.

In chapter 2, I will show how sophisticated customers, digital products, and crowdsourced data and technology sharing are driving enormous change in the marketplace. Customers are driving the generation of big data, empowered and incentivized by social media organizations, which offer free services to their users and use marketing and advertising as their funding model. The emergence of cloud computing has made it easier for marketers to source a lot more third-party data and use business partners to add additional value.

Chapter 3 discusses the first proposition—big data observations lead to an enriched customer profile. This chapter covers the first proposition as to how big data is significantly contributing to our ability to observe the consumer. It uses case studies to show how that factors into marketing research, customer segmentation, and delivery of marketing programs. It enumerates a number of big data sources, such as census data, social media, location data, web traffic, product usage data, and others. It also describes how marketers add value to the raw data using context and intent to drive customer insights.

Chapter 4 discusses the second proposition—automation and social media provide marketers with new ways of collaborative influence on the customer driven by personalized communication. This chapter describes how automation and social media are impacting our ability to communicate with customers, and these tools can be used to build collaboration across communities and build momentum for a brand. The success of a marketing organization is in its ability to strike a two-way communication, or a group communication in a community. A savvy marketer knows how to influence the creation of a buzz for a new product or a new message. In this chapter, I also discuss how customers' touchpoints have evolved to generate that communication on terms acceptable to customers.

Chapter 5 discusses the third and final proposition—marketing orchestration to optimize and customize a marketing plan for each customer or micro-segment. This chapter discusses how marketing dollars can be pooled across the silos to influence a customer through the stages of marketing. Market leaders selectively advertise based on the current state of the customer and effectively use promotions and expert testimonials to bring the consumer to a purchase decision. Orchestration also brings disparate organizations closer to each other to integrate and leverage data, insights, and actions across organizations.

Chapter 6 introduces the technological enablers. The changes in business functions are driven by a set of enablers. These capabilities have evolved significantly in the last five years and are driving many changes in how we market. Many books and blogs have defined big data using three, four, or six V's (velocity, volume, variety, veracity, and value, to name a few). In this chapter, I have restated those definitions as seen by the marketing community. Advanced analytics flourished in a big way toward semistructured and unstructured data. A number of machine learning techniques have replaced tedious manual processes for qualitative analytics. Experiment design has allowed marketers to convert their market into a gigantic lab for market experimentation, where product, pricing, and promotion decisions can be offered to smaller test markets and tested before global or regional launch. Customer identification and tracking techniques can help us identify specific customers or micro-segments across customer-generated events.

Chapter 7 shows how these changes are significantly changing marketing organizations—their metrics, processes, external business relationships, and people skills. This chapter looks at traditional marketing functions—marketing research, product management, advertising, pricing, media planning, promotions, and communication, and shows how these functions are being radically transformed by the three propositions. In addition, marketing organizations often employed shadow IT resources to conduct data integration, data mining, and visualization. These IT groups are increasingly investing into two roles—data scientists and data engineers. This chapter discusses how the analytical skills of the past, as practiced by marketing researchers, media planners, product managers, data modelers, database administrators, and marketing analysts, must be retooled.

The last chapter provides a focused view of corporate marketing and uses it as a special case to recount the main propositions in the book. Corporate marketing is, unfortunately, a comparatively under-researched area of marketing. However, it is going through its own transformation using big data, advanced analytics, and social media. The chapter outlines how corporate marketers are introducing these changes and how their changes are similar or dissimilar to consumer marketers.

I have tried to keep a narrative style in writing this book. One of my mentors in my early career days told me marketers are masters at storytelling and the best way to communicate with them is using stories. I hope you find the book informative and entertaining.

2

CHANGING WINDS

INTRODUCTION

We are creating more data each day than we created in entire decades or centuries in the past. How did we get here? Is it just a hype curve, an overfrothy bubble that will burst and then we will go back to good old small-sized structured data? Will social media keep buzzing? What if we do not find any way to make money from selling big data? Is there a crash coming? I have heard many doomsday scenarios about privacy, cost, overhyped sales promises, and boredom. Before we create a nice shiny picture of how marketing will be transformed by big data, it is important to examine the drivers for big data and the longevity of this tsunami.

In many ways, the foundation for big data was laid in the dot-com boom and bust. While some elements of big data may be overhyped today, the structure of society is fundamentally changing with big data. During the dot-com boom, telecom organizations bet on large capacities of dark fiber in anticipation of an all-electronic commerce vision. Since the dark fiber was available at an exorbitant sunk cost, the excessive supply of dark fiber crashed the costs of carrying bits from point A to point B. Storage companies invested heavily in faster and cheaper storage devices, leading to a major reduction of storage cost. The Internet for quite a while was the "World Wide Wait" as everyone rushed to mechanize their physical brochures, but then Google and Yahoo began to use their search engines to organize data. All in all, the dot-com boom and bust gave us the fundamental tools for big data, and told us the world had to be reengineered and not just mechanized.

Telecommunications executives often lament that their data transmission revenues have been capped due to an ultracompetitive marketplace, while the data carried by them has exponentially increased, leading to an unprecedented cost pressure on existing operations. Similar stories come out of media content and distribution companies—television producers, newspapers, magazines, and cable operators. In another way of viewing these changes, the combined market capitalization of Apple, Google, Amazon, and Facebook in 2013 exceeded $1 trillion, nearly all of it created in the last 20 years. Real-time bidding for online advertisements was unheard of ten years ago. Now it is more than half of online advertising business, which in turn is climbing to double digits in comparison with traditional advertising media as they lose their share of advertising revenue.

This chapter examines the market, social, and technological forces that have led to the data tsunami we see today. The chapter also examines the characteristics of big data and its use, and how it is different from the traditional data sources we have seen in the past. The lines of causality are never clear in retrospect. It may almost sound like a conspiracy theory, but we the marketers may have created the next wave. The time has come for the chief marketing officer (CMO) to come to the center stage.

So how are CMOs faring amid such turbulence? IBM conducted face-to-face interviews with 1,734 CMOs, spanning 19 industries and 64 countries, to find out what they are doing to help their enterprises cope with the fundamental shifts that are transforming business and the world. These interviews reveal that CMOs see four challenges as pervasive, universal game changers: the data explosion, social media, the proliferation of channels and devices, and shifting consumer demographics.[1] These challenges are driven by a set of environmental factors. In this chapter, I will examine these factors and show how they create both challenges and opportunities for marketers.

RISE OF DIGITAL SOCIETY

There are two common sources of data grouped under the banner of big data. First, we have a fair amount of data within the corporation that, thanks to automation and access, is increasingly shared. This includes emails; mainframe logs; blogs; PDF documents; business process events; and any other structured, unstructured, or semistructured data available inside the organization. Second, we are seeing a lot more data outside the organization, some available publicly, free of charge, some based on paid subscription, and the rest available selectively for specific business partners or customers. This includes information available on social media sites, product literature freely distributed by competitors, corporate customers' organization hierarchies, helpful hints available from third parties, and customer complaints posted on regulatory sites.

Sales and marketing have received their biggest boost in instrumentation from Internet-driven automation over the past ten years. Browsing, shopping, ordering, and receiving customer service on the Web have not only provided tremendous control to users but also created an enormous flood of big data into the marketing, product, and sales organization, enabling an improved understanding of buyer behavior. Each sequence of web clicks can be collected, collated, and analyzed for customer delight, puzzlement, dysphoria, or outright defection. More information can also be obtained about sequences leading up to a decision.

Self-service has crept in through a variety of means: Interactive Voice Responses (IVRs), kiosks, handheld devices, and many others. Each of these electronic means of communication acts like a gigantic pool of time-and-motion studies. We have data available on how many steps customers took, how many products they compared, and what attributes they focused on such as price, features, brand comparisons, recommendations, defects, and so on. Suppliers have gained enormous amounts of data from self-service and electronic sensors connected to products. If I use a two-way set-top box to watch television, the supplier has instant access to my channel-surfing behavior. Did I change the channel when an advertisement started? Did I turn the volume up or down when the jingle started to play? If I use the Internet to shop for a product, my click stream can be analyzed and used to study shopping behavior. How many products did I look at? Did I view the product description or the price when looking at the product? This enriched set of data allows us to analyze customer experience in the minutest detail.

Products are increasingly becoming digital. We read books and magazines on tablets, and listen to music on a variety of electronic media. These devices are capable of collecting a lot of statistics on consumer viewing or listening patterns. Apple iTunes keeps track of when I play each song and how often I play it. It can use this information to formulate sophisticated graphs showing how the various published music products are used and their affinity based on usage patterns. iTunes Radio or Pandora can offer a stream of new songs that are similar to the ones you just heard. Pandora lets you decide whether you like the new selection, and changes its selection based on your response. Unlike a typical radio station, which must rely on listener responses painstakingly collected through traditional means, these electronic radio stations can be rapidly tuned to custom music preferences. Amazon does a similar customization to book reading. Since the time I started to write my books and read a large number of books on my Kindle reader, Amazon has made my research easier by offering me new analytics books based on the books I am currently reading and publishing. Unlike my physical library, the Kindle library is providing me with valuable ways to share my books, and is offering Kindle ways to achieve a better understanding of collaboration among readers.

CONNECTED CARS

In the 1980s, as I was finishing my PhD at Carnegie Mellon University, I worked for a couple of years at the Robotics Institute and got my first exposure to autonomous vehicles. The thought of a robotically driven car that can navigate using a set of sensors and robots was at that time a novelty. Autonomous vehicles required a significant number of technologies in vision, navigation, and real-time decision-making. The work at the Robotics Institute led to an ambitious project conducted by the Federal Highway Administration, which involved a broader participation from a number of auto manufacturers and US universities.[2] Over the last two decades, we have seen a fair amount of progress toward making automobiles intelligent. Currently, a number of technology providers are teaming with automobile manufacturers to bring wireless technologies to cars in order to support the driver with safety, convenience, information, and entertainment.[3] While these manufacturers have yet to offer driverless cars commercially for the busy I-5 traffic in Southern California, they provide a number of safety and convenience features in cars today using the underlying technologies.

One of the important developments is in the instrumentation of the car. Today's cars come equipped with wireless and sensor technologies to keep track of all the mechanical and electrical components. Instead of using expensive instrumentation at the dealer's service center, which would not be able to collect performance information unless the failure conditions are replicated, today's cars use sensors to collect the data as we use our cars. This data can be used for sophisticated analytics once we take the car to the dealer. However, if the car is equipped with Internet connectivity, it can be monitored remotely. As faults or other service requirements are detected, the dealers or manufacturers can provide intelligent alerts to the driver, before a fault leads to a major problem. Once the data has been tapped from the vehicle, it can be used for a variety of purposes. For example, car insurance companies can use this data to provide insurance discounts to good drivers.

Cars provide marketers enormous opportunities for location-based advertising. As drivers increasingly use gadgets to plan and record their travel, a marketing agent can observe, analyze, and recommend activities. As I start running out of fuel, it can present options for refueling. After driving for a while on a long journey, it can offer rest stops, places to eat, and shop. In a new city, it can be a concierge as a family looks for activities. Navigation products have already started offering products in this area. However, marketers have barely scratched the surface in using automobile data or gadgets for understanding or communicating with consumers.

UBIQUITOUS USE OF MOBILE PLATFORM

The first documented evidence on wireless phones dates as far back as 1908, when Professor Albert Jahnke built the first device, which communicated without wires in Kansas, Missouri.[4] While wireless radio technology was widespread during World War II, Motorola first introduced handheld wireless devices in 1973.[5] However, the last 20 years have seen an unprecedented growth in mobile devices around the globe. Now, handheld mobile devices are the primary mode of communication in many growth markets, surpassing landline phone systems.

Smartphones and tablets have rapidly engrossed customers in usages other than voice communication and attracted attention from marketers. I am fascinated with their increasing use as a universal device for everything—a remote control for changing television channels, a shopping device at the store, a music store, a stop watch, an alarm clock, and so on. The other day, my neighbor was using her iPhone mirror app to adjust her makeup. However, initial marketing on wireless devices was fairly primitive—using short message service (SMS) or display advertising on the tiny screen. With the maturing wireless space, marketing budgets and techniques have exploded on the mobile platform. Marketers are placing a clear premium on reaching audiences via mobile device screens in the future. While 86 percent of respondents said mobile phones were currently important advertising channels, 98 percent said they expected them to be important in three years' time.[6] Tiny display advertisements are no longer the norm. As I watch YouTube on my tablet, the site gives me the option to skip the advertisement after the first ten seconds of viewing. However, if I continue to watch, the video may go on for a long time, in many cases for minutes.

Mobile platforms have invaded the shopping process and now provide great options for additional information access at shops. These location-aware phones can be pinpointed within 20 yards of accuracy in a shop offering Wi-Fi service, and marketers have provided additional information about products, which can be accessed on the phone. Shoppers also use their phones for comparative shopping, seeking advice, comparing prices, and looking for deals from the competition. In response, stores are actively using phones for location-aware marketing, mobile coupons, loyalty cards, and related marketing activities. Since a mobile device is location aware, it can be used for active placement of location-aware marketing messaging. Also, companies are using mobile devices to develop interactive marketing programs. For example, Foursquare (www.foursquare.com) encourages me to document my visits to a set of businesses it advertises. It provides me with points for each visit and rewards me with the title of "Mayor" if I am the most frequent visitor to a specific business location. Every time I visit Tokyo Joe's, my favorite neighborhood nearby sushi place, I let Foursquare know about my visit and collect award points. Presumably, Foursquare, Tokyo Joe's, and all the competing sushi restaurants can use this information to attract my attention at the next dining opportunity.

Online commerce in the United States keeps growing each year on Black Friday (the day after Thanksgiving, a major US holiday). Traditionally, shoppers have lined up in front of brick-and-mortar stores on the Friday after Thanksgiving. However, mobile devices have provided additional shopping and buying opportunities, and online retailers have been gearing up for Black Friday sales upticks. IBM used its cloud-based analytics on 2013 Black Friday shopper behavior to report that mobile traffic grew to 31.7 percent of all online traffic, increasing by 45 percent over 2012. Mobile sales were also very strong, exceeding

17 percent of total online sales, an increase of 55.4 percent year-over-year. Smartphones drove 19.7 percent of all online traffic, compared to tablets at 11.5 percent, making them the browsing device of choice. When it comes to making the sale, tablets drove 11.7 percent of all online sales, more than double that of smartphones, which accounted for 5.5 percent.[7]

In addition to providing communication and content, mobile platforms are rapidly gaining popularity for commerce. Mobile wallets can be used for carrying digital money, which can be used to buy goods. In developing countries, where banks are relatively scarce, mobile wallets are gradually becoming quasi banks, as they are used not only for the buying and selling of goods but also for lending money. Like credit cards, mobile wallets can be used instead of money and record the place of transaction. Coupled with communication and content information, they can also provide a fair amount of shopping history before the purchase is made.

The biggest change in the use of mobile technologies is coming from the large number of apps available for both consumer and corporate computing. According to Appbrain, there were 879,506 apps on the Android market as of December 8, 2013.[8] Apple has traditionally been a little ahead, and it crossed the million apps mark in October 2013.[9] Apps are increasingly transforming mobile devices into enriched browsing and interaction platforms. Many of these apps are location aware and able to provide location-specific capabilities after specific permissions have been obtained from the users.

SAVVY CUSTOMERS DISCOVER SOCIAL NETWORKING

As I watch my two kids shop, I feel envious of how they make their decisions today compared with what I did 35 years ago. The increase in electronic collaboration has created a new breed of sophisticated consumers. These consumers are far more analytical, far savvier at using statistics, and far more dexterous at seeking collaboration from near-strangers to rapidly collect and collate opinions from others.

I have been leisure traveling extensively for over three decades, and it took me months each time to plan the vacation. Thirty years ago, my usual process of searching for the best places to visit and restaurants at which to eat was based on American Automobile Association (AAA) brochures. As the Internet gave me the flexibility to search destinations online, I mechanized my searches, without significantly changing the process. I used to scour the Internet looking for brochures, read their descriptions, and guess which restaurants would be closest to our liking. In these years, as my kids grew, they laughed at my archaic ways of shopping. In a spirit of their showing me how it works, I finally gave up control over our travel to Italy for a family vacation in 2007. My kids were living at opposite ends of the United States, but were able to collaborate with each other, their friends, and others to organize us, using a new set of tools, without paying a single cent to AAA for membership or Barnes & Noble for expensive travel books. They went through sites like Yelp and TripAdvisor and sought help from their friends (online, obviously) to find rare places that were both inexpensive and served delicious food. We were never lost for directions, always found ethnic foods I could never have located through my old ways, and paid a fraction of what I would have paid for food. We ate almost like locals in Rome and Florence, and by throwing a couple of Italian phrases into the mix, we almost acted like ones, too. Yet while I admired my kids' way of working, it is relatively harder for me to adopt their style of travel planning. I was organizing a trip to Rajasthan, India, in December 2013, and my gut reactions were still the same. My daughter, who had never been to Rajasthan, had already figured out the best restaurants in Jaipur, the mode of transportation to the Taj Mahal, and the best tourist spots. The comparisons were provided not by marketers, but by other travelers like her.

The information collected by travelers was organized at a fraction of the cost of collating, organizing, and publishing travel knowledge in proprietary ways. It was the result of a gigantic collaboration. Some of the data was very biased and very incorrect. However the data tsunami took care of the biases. My daughter explained to me that the true value of high scores was only valid when there was large number of recommendations. "If a restaurant has four or five stars with less than ten reviewers, it may be fabricated reviews from family and friends," she said. "You should only go for recommendations where the number of reviews is large. And, do read the reviews, as there is a lot more data hidden in the unstructured text," she added. While this was common sense, it was interesting to see how social media sites had put that information in the hands of consumers on their devices. It was a lot of unstructured data, but it gave us much more than simple quantitative scores. Not everyone was looking for the same product, but by reading the reviews, we could get so much more information.

How did so many people collaborate so effortlessly in Yelp? How is the data organized so easily and so well? How do we have a taxonomy in which I can type any small ethnic community food type and find half a dozen restaurants in Southern California? These sites are mostly crowd-sourced, with strong governance and a well-maintained taxonomy that is changed based on how the users of the community search. Yelp has a way of rewarding its best reviewers. They are called elite reviewers:

The Yelp Elite Squad is our way of recognizing and rewarding yelpers who are active evangelists and role models, both on and off the site. Elite-worthiness is based on a number of things, including well-written reviews, great tips on mobile, a fleshed-out personal profile, an active voting and complimenting record, and playing nice with others. Members of the Elite Squad are designated by a shiny Elite badge on their account profile.[10]

Today's customers can shop around the globe, find out more than ever before about the organizations they are dealing with, and share their views with hundreds of thousands, if not millions, of fellow customers. Their expectations—be they consumers, citizens, or business customers—are soaring. And they can make or break brands overnight!

Society has always played a major role in our evaluation process. However, the Internet and social networking have radically altered our access to information. I may choose to "like" a product on Facebook, and my network now has instant access to this action. If I consider a restaurant worth the money, Yelp can help me broadcast that fact worldwide. If I hate the new cell phone service from a telco, I can blog to complain about it to everyone.

We have to be careful in using these powerful tools on big data, however, as they are likely to reveal private facts. The *Wall Street Journal* reported the case of a student who joined a gay community. The community, in turn, invited her to join a Facebook page where the community was posting its events and sharing friends.[11] Unfortunately, this student had not disclosed her gay status to her parents. To her shock, an invitation went from her community to her entire friends circle, including her parents, to join the community. As Facebook fine-tunes its privacy preferences and policies for posting and broadcasting information, such stories are grim reminders of the complexity in our social structure and how social media must weave its designs to meet these requirements.

On a positive note, Barack Obama's presidential campaign is a great example of how a political campaign started at the grass-roots level, connected with its constituencies, respected them for their individual differences, and focused a multichannel campaign using a gigantic voter data model. The campaign built a detailed data model and kept track of individual political views and preferences, encouraging participation at the grass-roots level and constantly using these activities to update the data model. The campaign then aggregated the data from the individual level to formulate clusters and directed the rest of the channels, including the venue for President Bill Clinton's speeches and television spots, to align with the pockets of voters requiring the most attention. The email campaign continued even after the election. As a participant, I receive well-written emails from the president himself or the First Lady, sharing with me his activities that pertain to my wish list of issues and encouraging me to participate in the political process by continuing to voice my opinion. The big data analytics for the campaign were choreographed by Dan Wagner, and have been extensively discussed by political and technical journals.

The significance of Wagner's achievement went far beyond his ability to declare winners months before Election Day. His approach amounted to a decisive break with twentieth-century tools for tracking public opinion, which revolved around quarantining small samples that could be treated as representative of the whole. Wagner emerged from a cadre of analysts who thought of voters as individuals and worked to aggregate projections about their opinions and behavior until they revealed a composite picture of everyone.[12]

CROWDSOURCED ANALYTICS TOOLS

My son and I have learned analytics across two generations. I think I belong to the second generation of computer scientists, as I grew up with personal computers and did not invest any of my time in the IBM 360-class machines. My son grew up with the Internet and the open-source community. As you can well imagine, he has typical Silicon Valley gut reactions, while I am still trying to weigh-in with my DNA from corporate America. We had an interesting discussion regarding his algorithms. I suggested that he should file a patent for them so that he can protect his intellectual properties. However, he was more interested in contributing them to an open-source community, so he could get others to work with him.

Although open-source software sources have been around since much of the 1980s and 1990s, their popularity grew exponentially with the World Wide Web. The "open-source" label was created at a strategy session held on February 3, 1998, in Palo Alto, California, shortly after the announcement of the release of the Netscape source code. The strategy session grew from the realization that the attention around the Netscape announcement had created an opportunity to educate and advocate for the superiority of an open development process. The conferees believed that the pragmatic, business-case grounds that had motivated Netscape to release their code illustrated a valuable way to engage with potential software users and developers, and convince them to create and improve source code by active participation in a community. The conferees also believed that it would be useful to have a single label that identified this approach and distinguished it from the philosophically and politically focused label "free software." Brainstorming for this new label eventually converged on the term "open source," originally suggested by Christine Peterson, co-founder of Foresight Institute.[13]

The Apache site owes its genesis to the HTTP daemon. In February 1995, the most popular server software on the Web was the public domain HTTP daemon, developed by Rob McCool at the National Center for Supercomputing Applications (NCSA), University of Illinois, Urbana-Champaign. However, development of that httpd stalled after McCool left NCSA in mid-1994, and many webmasters developed their own extensions and bug fixes that were in need of a common distribution. A small group of these webmasters, contacted via private email, gathered together for the purpose of coordinating their changes (in the form of "patches"). Brian Behlendorf and Cliff Skolnick put together a mailing list and shared information space and logins for the core developers on a machine, with bandwidth donated by HotWired.[14] As of June 2013, Apache was estimated to serve 54.2 percent of all active websites and 53.3 percent of the top servers across all domains.[15]

Google released the Google File System paper in October 2003 and the MapReduce paper in December 2004, which attracted the attention of Doug Cutting and Mike Cafarella at the University of Washington, who were developing the Nutch, an open-source search engine. In 2006, Cutting went to work at Yahoo, which was equally impressed by the Google File System and MapReduce papers and wanted to build open-source technologies based on them. They spun out the storage and processing parts of Nutch to form Hadoop (named after Cutting's son's stuffed elephant) as an open-source Apache Software Foundation. However, although Yahoo was responsible for the vast majority of development during its formative years, Hadoop did not exist in a bubble inside Yahoo's headquarters. It was a full-on Apache project that attracted users and contributors from around the world.[16]

Hadoop gained the attention and mindshare of a large number of developers globally who used the Apache server to share their ideas and code. Since the entire program is open source, it is now gathering a fair amount of momentum at corporate information technology (IT) organizations, and has risen as a serious competitive threat to the traditional software for data storage, integration, and analytics. It has attracted a large technical community, which is contributing to the development of software. In addition, the academic community has created a large number of online courses.

As much as data about marketers was crowdsourced by social media organizations, much of the storage and analytics techniques are also crowdsourced. It has been a fascinating experiment of public sharing, which has attracted market leaders, such as IBM, who are contributing and benefiting from this open-source technology development program.

Unlike a patent, which protects ideas but creates pockets of innovations requiring expensive integration (some of which are also patented), the open-source experimentation allows a developer to share initial ideas and use a community to improve the original idea. Each developer has access to the source code and can use it to create commercial products, which, through the very nature of their development, are far better integrated.

MONETIZATION

The music industry has seen its revenue models turned upside down with digital music availability and piracy. So, how would a musician popularize his/her songs and monetize the song? Let me take an example from Bollywood, where free downloads shattered the traditional music business. Before the invasion of the Internet and YouTube, the marketing of songs in India was tied to movie releases, and therefore the success of a song was closely tied to the success of the movie. As a movie was released, hundreds of millions of moviegoers evaluated the fate of the song. Often, the songs carried the movie or never rose to popularity because the movie flopped and most of the potential audience did not get a chance to hear the song. However, with the rise of the Internet, social media, and digital recordings, the Indian music industry is going through a cataclysmic change.

The "Kolaveri" song by Dhanush presents an interesting story of how Twitter, Facebook, and YouTube became its marketing and monetization instruments. The song was written in Tanglish (a combination of Tamil and English) and leaked in an early recording. With its poetic lyrics, catchy beat, and the informal nature of its launch, the leaked video rapidly became a marketing sensation. At that time, the music publishers were not even considering YouTube as a possible publication medium. Yet due to the song's viral success, they decided to officially launch it on You Tube on November 16, 2011. It might seem that Sony acted fast, sealing the deal within 13 days of the video's upload. But by then, the song had already clocked 9 million views. Had Sony managed to monetize those views, it would have made an additional $4,000 (Rs 200,000 at the prevailing rates in 2011), based on YouTube's $1 cost per 1,000 impressions (CPM) assuming a 50 percent revenue partnership. The song rapidly became a big sensation among youth in India and recorded the highest number of Twitter hits in India in record time. The song received an unusual amount of attention from celebrity movie stars and singers. It was Facebook that emerged as the main driver for "Kolaveri," accounting for 80 percent of social media mentions, followed by Twitter and YouTube, according to Social Hues.[17] Sony has since been collecting revenues through the YouTube views, which totaled over 70 million by fall 2013.[18] The song was created for the movie "3," which, unfortunately, did not succeed at the box office.[19] Had the song used the movie to promote it, as the Indian movie and music industry had done for so many decades, its success would have been tied to the success or failure of the movie.

"Kolaveri" is nowhere close to the top of viewed YouTube videos, however. The video "Gangam Style" from the Korean group Psy was the first to hit 1 billion YouTube views.[20] Television viewership is actually declining among teens and youth. YouTube and other online channels have snatched market share from the conventional media. While many of the big brands are rapidly discovering YouTube as a publication medium, it has also spawned a large number of amateur content providers, and is used as well for sharing videos among family and friends. Content providers are able to employ YouTube as a platform for publication and advertising. It has also prompted entrepreneurs like Lisa Irby to teach others how to monetize using Google Adsense.[21]

Google took an early lead in monetizing the online search with its innovative clicks-based revenue model. It rapidly led to a social media monetization program across many sites. However, many social media communities have remained advertising free and do not have a goal of making profits, for example, Wikipedia and Craigslist.

So far, we have discussed advertising revenues from content distribution. But how about storing and analyzing customer data that comes from content viewership? From a big data analytics perspective, a "data bazaar" is the biggest enabler to create an external marketplace, where we collect, exchange, and sell customer information. We are seeing a new trend in the marketplace, in which customer experience from one industry is anonymized, packaged, and sold to other industries. Fortunately, Internet advertising came to our rescue in providing an incentive to customers through free services and across-the-board opt-ins.

Internet advertising is a remarkably complex field. With over $20 billion of revenue in the first half of 2013[22], the industry is feeding a fair amount of start-up and initial public offering (IPO) activity. What is interesting is that this advertising money is enhancing the customer experience. Take the case of Yelp, which lets consumers share their experiences regarding restaurants, shopping, nightlife, beauty spas, coffee and tea, and so forth. Yelp obtains its revenues through advertising on its website; however, most of the traffic is from people who access Yelp to read customer experiences posted by others. With all this traffic coming to the Internet, the questions that arise are, how is this Internet usage experience captured and packaged, and how are advertisements traded among advertisers and publishers.

Terence Kawaja has been studying this market since 2009, when he created a chart to explain the emerging online advertising market. To date, slide versions of his chart have received more than 350,000 views online, from people in 116 countries. Sure, compared to YouTube videos of cute babies and cats that rack up millions of views in a matter of days, 350,000 is quite small. But given the obscure nature of the material (a chart full of ad tech acronyms such as "SSPs," "DSPs," and "DMPs"), it is a fairly impressive number.[23] I have used his charts in my books. In most of my presentations, I could measure the maturity of the audience by the number of LUMAscape acronyms they were familiar with. "Terence Kawaja has a new way for potential investors to visualize it," says *Wall Street Journal* writer Amir Efrati. "The market involves hundreds of small and large companies that help advertisers reach consumers and help web site publishers, mobile-application developers, search engines, and other digital destinations generate revenue through advertising."[24] Over the years, Kawaja has added different charts to represent various markets—display, video, search engines, mobile, social, and commerce. For the latest LUMAscapes, visit Kawaja's web site: *www.lumapartners.com*. A number of intermediaries play key roles in developing an advertising inventory, auctioning the inventory to the ad servers, and facilitating the related payment process, as the advertisements are clicked and related buying decisions are tracked. I have been following the LUMAscapes for over two years. A number of my clients have mentioned LUMAscape as their source of data for finding companies to work with. In chapter 4, I will describe the mechanics and benefits of real-time bidding in more detail, and in chapter 7, I will discuss further the roles of different players and how they participate in the advertising process.

In addition to Internet advertising and usage data, a number of other markets are rapidly emerging for data monetization. Telecom companies have been experimenting with the monetization of location and device usage information. However, a couple of regulatory and customer privacy preference concerns have kept them in check. For a typical telecom provider, the potential revenue from monetization is insignificant in comparison to the revenue from telecom usage. A backlash could potentially set them back in comparison to the competition. In addition, various regulations hold them responsible for customer data privacy.

PRIVATE AND PUBLIC CLOUD

Thus far, I have covered social media as a source for data, open source as the mechanism for finding software tools, and monetization as the motivation for burning the midnight oil. But how can I have the computing environment for such large data storage and analysis? Cloud technology has offered an attractive proposition to anyone who wants to experiment with marketing analytics and lacks resources. I met a group of students at Stanford University who were conducting a sentiment analytics of Twitter data in the early days of unstructured data analytics of big data. They were dealing with terabytes of data and running sophisticated algorithms to interpret unstructured data and seek Twitter influence with positive and negative sentiments.[25] They told me they used Amazon cloud to perform this extensive and valuable analysis on a tight budget without having to buy and build a dedicated infrastructure.

Time-shared computing was popularized by mainframe computers, which were expensive to buy and house. Getting leased access to the mainframe was the best way of using it without incurring massive capital and other fixed costs. Clouds apply the same principle to the modern computing infrastructure. A variety of cloud solutions have emerged over the last decade. Public clouds house data across many corporations or consumers and offer secure access to each. Private clouds house the data within a corporation's firewalls, but use the cloud infrastructure to reconfigure the environment for each project, thereby reducing the dedicated purchase of computing infrastructure for each project. Cloud technology can be used at different levels of computing infrastructure. An infrastructure cloud offers a computing environment. A storage cloud offers capabilities for storing data as well as capabilities for backup and restore. For example, Symantec and Amazon provide storage clouds for personal computer data backup. An application cloud houses an entire application, such as SalesForce.com.

As large quantities of public data started to emerge, cloud providers offered ready-to-use solutions for analysis of this data. It is an easy decision to use a public storage for already public data and analyze it for specific queries. Cloud providers offered low entry points and subscription fees to simplify starting costs for analytics. As offerings from cloud-based analytics, such as Coremetrics[26], Radian6[27], and Attensity[28] proliferated, marketers began to augment their traditional sources and manual scans of public data using public cloud analytics sources as input. In a study with a large telco, I found a number of subscriptions to competing cloud-based providers, some of them overlapping in capabilities. The price tags were reasonable, and the starting costs were almost negligible.

The pricing models for cloud-based analytics providers have significantly challenged the software and services industry. In a typical capital purchase, software is sold with an upfront fee and an annual maintenance fee, and customization services are also front loaded and may cost as much as or more than the software. For a typical marketing automation program, nearly 50 percent of the cost may need to be incurred in the first year, while most of the benefits may be back loaded. To make these programs viable, most of the large programs were capitalized with a multiyear amortization schedule. The cloud changed the model to a monthly subscription, where the costs are mostly transaction driven and, hence, back loaded, while the benefits may be accelerated through early deployments. In addition, the marketer may choose to cancel the program any time without incurring expensive upfront costs. Information Management reported a story on Adobe management's decision in August 2011 to dump its lucrative licensing business in favor of a monthly subscription offering called Creative Cloud. The same shift is underway across the technology industry as vendors vie for a bigger slice of the $36.8 billion cloud software market, which will balloon to $67.3 billion by 2016, according to IDC.[29]

As usage grew, marketers began to raise data security and ownership questions. The market is now maturing rapidly into using a hybrid environment, where public data is being analyzed using public cloud offerings, but its correlation with internal marketing data is conducted in a private cloud. Private clouds provide many of the benefits of public clouds, and yet offer the benefits of data and ownership protection.

CUSTOMER PREFERENCES AND PRIVACY CONCERNS

Data privacy for big data is a serious business, and is causing regulators around the globe to set up a variety of policies and procedures. We have witnessed debates regarding the use of location data for monetization as well as other commercial and government uses. The US Federal Trade Commission settled a case with Facebook that now requires the company to conduct regular audits. Facebook, Inc. agreed to submit to government audits of its privacy practices every other year for the next two decades. The company also agreed to obtain explicit approval from users before changing the type of content it makes public.[30] Similar processes have been put in place at MySpace and Google. Under certain opt-in conditions, the collection and storage of location information may be permissible. Also, some of the data can be made anonymous and used for statistical analysis. Two bills introduced in the US House of Representatives and the US Senate limit how the government and private companies can use information about customer location. Both bills await approval by the House. The bills are among multiple efforts in Washington, DC, to update digital-privacy laws, particularly as they relate to location. One bill, sponsored by Democratic Senators

Al Franken of Minnesota and Richard Blumenthal of Connecticut, requires companies such as Apple and Google, as well as the makers of applications that run on their devices, to obtain user consent before sharing information with outsiders about the location of a mobile device. The other bill, by Senator Ron Wyden (D., Oregon) and Representative Jason Chaffetz (R., Utah), requires law enforcement agencies to obtain a warrant in order to track an individual's location through a mobile phone or a special tracking device. This follows an earlier bill introduced by Senate Judiciary Committee Chairman Patrick Leahy (D., Vermont) that imposed a similar requirement and also required law enforcement to obtain a search warrant in order to retrieve old emails stored on servers. The latest public relations uproar over the use of location data by the National Security Agency (NSA) has created a fair amount of public clamor.[31] The laws concerning when the government can track someone's location are murky. One key law dates from 1986, before the widespread use of cell phones or global positioning satellites.

At the end of the day, privacy preferences are personal to consumers and vary widely among individuals. In many situations, marketers have found good reception for the use of customer data in exchange for a service. In many cases, consumers will trade their privacy for favors. For example, my cable/satellite provider sought to have my channel click information shared with a search engine provider. They offered me a discount of $10 if I would "opt-in" and let them monetize my channel-surfing behavior. An early example was Apple's use of device data for product improvement in exchange for its "locate iPhone" service. Based on consumer reaction, Apple fine-tuned its privacy preferences and clarified that it collects location data only as needed for improving location services for Apple and its business partners[32] Sprint and CarrierIQ tried an early step toward using device and network analytics and found stiff opposition to their plans.[33] In May 2013, AT&T revamped its privacy policy in which aggregate location data can be used for delivering customized content, relevant advertising, and personalized offers for products and services that may be of interest to customers.[34] Consumers have the opportunity to opt-out. Privacy laws also differ across geographies. A global marketing program using customer data for marketing must adjust to regional laws and preferences.

This leads us to several interesting possibilities. Let us say that a data scientist uses the channel-surfing information from cable viewing to characterize a household as interested in sports cars (for example, through the number of hours logged watching NASCAR). The search engine then places a number of sports car advertisements on the web browser used by the desktop in that household and places a web cookie on the desktop to remind them of this segmentation. Next, a couple of car dealers pick up this "semi-public" web cookie from the web browser and manage to link this information to a home phone number. It would be catastrophic if these dealers were to start calling the home phone to offer car promotions. When I originally opted-in, what did I agree to opt-in to? And is my cable/satellite provider protecting me from the misuse of that data? As we move from free search engines to free emails to discounted phones to discounted installation services, all based on the monetization of data and advertising revenue, there is money for everyone if the data is properly protected against unauthorized use. Many technology providers are selling data obfuscation processes to protect customer privacy. Most of the time, marketers are interested in customer characteristics that can be provided without privately identifiable information (PII)—that is, uniquely identifiable information about the individual that can be used to identify, locate, and contact that individual. We can possibly destroy all PII, which may still provide useful information to a marketer about a group of individuals. Now, under "opt-in," the PII can be released to a select few as long as it is protected from the rest. In the preceding example, by collecting $10, I may give permission to a web search engine to increase sports car advertisements to everyone in my Zip+4, while at the same time expecting protection from dealer calls, which require a household-level granularity. We can provide this level of obfuscation by destroying PII for house numbers and street names, while leaving Zip+4 information in the monetized data.

I have banking/investment accounts with five major financial institutions. One of these banks recently approached me about consolidating all my bank accounts with them. As we were going through the details, I was being asked to share a fair amount of private information. I wondered how much the bank already knew about me since I have dealt with them for over a decade and given them access to credit reports and mortgage applications. Also, a data scientist at the bank could correlate information authorized by me, information that is publicly available, and self-provided personal information. How is this full and complete view of my customer profile stored and accessed at the bank?

We have heard about data security breaches. Recently, the *Wall Street Journal* published an article about a Yahoo! security breach that exposed 453,000 unencrypted user names and passwords.[35] Is all this data that the bank is collecting about me safe? Often, we assume a large global brand is safe; however, the recent data breaches include a long list of famous brand names.

The technology is continuing to evolve in their use of web tracking information. Since most of the information sharing on the Internet is stateless, "web cookies" grew as informal mechanisms to track a user of a website. Unfortunately, cookie data was not well protected, and a number of companies began to analyze cookies to look for private information. Cookies are also not as reliable source of data, as they are often deleted and are not used by the mobile platforms. Other ways of tracking customer data, such as fingerprinting the device are being developed.[36]

Data privacy is an important concern, and is often discussed in the context of consumer targeting. Even if I suppress PII, I may target an individual or a group of individuals with specific marketing messages that may be considered to be violation of privacy by the customer. How do we draw the line, and how do we still converse with customers? It is all about customer context and trust. As customers, we trust some relationships and are willing to open up. As long as the trust is maintained, information or marketing propositions can be freely exchanged. However, if the trust is broken, it creates irreparable damage not only for that relationship but also for many other relationships. It is like a marital relationship. A couple comes together with a "trust" that lets them relate to each other. If the trust is broken, it impacts not just the couple's, but many other relationships. As marketers, we owe it to our brands as well as to the rest of the marketing community to build and maintain trust with customers, which allows the exchange of information and marketing offers.

SO HOW DOES IT IMPACT MARKETING?

While we talked about the "voice of the customer" and the "customer is king" even before the Internet and social media became popular, the communication traveled from marketers to customers. It was like a cable modem, through which downloads could be executed at broadband speeds, while uploads were tiny narrow bands of sample data. However, communication has now all changed and unleashed the biggest worldwide chatterbox! The megaphones are now all in place, and we just need a little instigation. However, this is broadband in reverse. Communication from customers is no longer restricted. It may overwhelm our listening devices as well as our warehouses, and those of us who ignore it, may get trampled over by those who master it.

The spirit of Arpanet as it evolved through research institutes and universities was to preserve the independence and spirit of altruistic cooperation without commercialization. Internet amplified that spirit, and social media added a lot of voices to it. For the longest time, Facebook remained commercial free. The intent was once again to build an environment of altruistic collaboration. Yet, thanks to marketers and the pressure for public performance, commercialization is getting introduced. However, commercialization faces a large population of users who collaborated in altruistic ways in the past. Marketers can benefit from this crowdsourced participation as long as they follow the spirit.

As mobile platforms grew, marketers started using them for marketing. Although it has taken us a while to understand the real power of mobility, it is all about understanding consumers and their behaviors. It is probably the richest source of data, and is a gold mine for understanding the context and intent of our customers.

These major trends must be harvested using a marketing strategy that revolves around collaboration with customers. Customers are ready to work with marketers as long as marketers build trust and engage them in a productive dialogue that benefits the customers. These trends obviously support marketers, too, as they get to sell their products. However, these trends shift the balance of power to the customer and give them the ability to influence our products, our messaging, and our relationship with customers.

The power remains with the customers and the impartial review process. Amazon reviewers can tear apart a bad product, and if fake reviewers join in to incorrectly praise a product, they get punished along with the product. Yes, we can include celebrity endorsements, and for a fee, they may add a voice, which is no different from classic Marketing 101. However, a digitally connected society has its checks and balances to fairly evaluate a product. When savvy customers shop for products, they are dexterous at weighing paid celebrity endorsements against the voice of the masses.

3

FROM SAMPLE TO POPULATION

INTRODUCTION

During my last trip to Mumbai, India, I was staying in a hotel at their famous Juhu beach. My trip was relatively short, and I was debating whether to indulge in social activities. In the early morning, I was tempted to go for a jog at the beach. By the time I sat down for breakfast, I had messages from two of my friends asking how long I was going to be in Mumbai. I track my jogging activities on Endomondo, which is linked to my Facebook page. I did not have to call any of my social circles to announce my arrival in Mumbai, as Endomondo took care of it.

Big data is beginning to provide us a powerful mechanism to record observations in a public way and broadcast them worldwide, at Internet speed. Each photo I place on Facebook is available to my social circles. I come from a big family and have 65+ first cousins distributed across many countries and continents. As I immigrated to United States in early 1980s, I started to lose contact information and other pertinent details. As my family and I graduated from sharing addresses and phones to emails and Facebook pages, we found it was relatively easier to renew connections with the help of digital communication, and social media, as most of my cousins had email addresses and were active on Facebook. Although today the extended family is distributed all over the world, we feel connected to each other, as we see each other's activities, share in commenting and placing "Like" buttons, and can connect to each other despite each of us constantly changing contact information.

So, what are these observations? How do we participate in these observations? Do consumers have the ability to control how to make these observations, private, public, or shared with a select group? How biased are these observations? As marketers, can we rely on these observations? How do marketers convert observed data into meaningful insight? Where are the limits to the computing power in dealing with velocity, volume, or variety of data? This chapter will tackle these questions and position the first proposition to marketers.

CENSUS DATA

Official nationwide censuses have been the original big data sources for centuries. Censuses were conducted across a number of ancient societies, including China, India, Egypt, Greece, and Rome. According to *Webster's* dictionary, a census is the official process of counting the number of people in a country, city, or town and collecting information about them. A census is an expensive and time-consuming proposition. Most of the ancient kingdoms used a census to keep track of the population driven by their political needs—primarily the need to keep accurate count of ethnic communities as well as for taxation. The US Constitution has stipulated that a census be conducted every ten years, and by 1880 the size and geographical breadth of the US population drove the manual process to its limits, fueling automation and the use of punch-card machines.[1] Even today, the primary beneficiary of the census in the United States is the congressional district zoning process. However, it remains a costly and time-consuming process, as each citizen must respond to a questionnaire with a large number of questions. For a country of more than 300 million persons spread out over 3.6 million square miles, counting the entire populace is an immense logistical feat. To accomplish it for the 2010 census, the

Census Bureau mailed approximately 134 million questionnaires that were to be completed by April 1. That would cost nearly $60 million in postage alone if the Census Bureau did not get free postage from the United States Postal Service (USPS). The collective weight of all 360 million printed questionnaires (from all three mailings) is nearly 12 million pounds, and if stacked on top of one another, would be nearly 29 miles high.[2] Electoral candidates as well as pollsters extensively used census data in building their big-data-driven prediction models for election results. As I watched the counting of votes in the 2012 US elections, I was fascinated with how John King from CNN used the census data and his predictive model to provide early analysis of the results.[3]

While a census is mandated by the constitution in many nations around the world as an important input for organizing the democratic division of the voting process, it provides a wealth of information to the marketing community. It is by far the most comprehensive view of a nation, and by combining data across many nations, a marketing organization can collect a global view of its consumers. Marketers have mastered the art of combining statistical information collected from small samples and projected to the entire population using census data. For example, Nielsen's report on Asian consumers uses census data to accurately estimate the size of the Asian population in the United States and then employs a large number of statistically significant samples of the population to project the behavior of these consumers.[4]

A census is also an important case study on the protection of personal data. While the data is collected systematically across the entire population, from each individual, its public access is typically in the form of aggregate data. All the collected data is available at an appropriate aggregate level that does not reveal the identity of an individual, while at the same time it provides valuable information about a community.

Statistical sampling offers us with an important way to collect detailed data from a small number of people and achieve a relatively

high accuracy in our ability to predict the behavior of an entire population, as long as the sampling is done "randomly," that is, each individual
has equal probability of being chosen as a sample representative of the
population. For example, if we were to conduct a telephone survey to
elicit opinions in a society where only the wealthiest 10 percent of individuals use telephones, the sample would not be random. For a long
time, with a census as the only source of big data, there was no easy way
to challenge a prediction based on statistical sample data projected to
the population using census data. In the recent past, we began to see
other sources of big data, which are an extensive representation of the
society at large. In many cases, they represent observations as opposed
to reported information. How do we combine census data with these
other sources to analyze and infer consumer behavior? Let me discuss
a couple of examples.

SOCIAL MEDIA DATA

Most of the early marketing analytics efforts were initiated from studies of unstructured texts from public websites. There are two strong
reasons why this data provided the fuel for early research. First, the
data on public websites is easily available to the academic community.
While the early work started with research publications and search
engines, it began to branch off rapidly into many aspects of marketing
analytics. Second, the data provided an enormous richness to develop
powerful hypotheses on social behaviors, which could be studied and
perfected. Unlike structured data, the analysis of unstructured text has
many nuances. Researchers have to build techniques for understanding and translating human writing, for example, converting unstructured data into structured sentiment scores. A blog or tweet may
carry sarcasm or emoticons, which should be interpreted by computer
programs.[5]

   Jonathan Taplin has been a seasoned businessman in the entertainment business, having produced successful concerts and movies. He is
now the director of the University of Southern California's Annenberg

Innovation Lab and has been working closely with IBM in studying how unstructured data can be analyzed for Hollywood Marketing.

> *Our results demonstrate not just the usefulness of monitoring social sentiment but the importance of deeply analyzing the raw results so marketing leaders come away with a precise understanding of what consumers think and want. For example, before the mid-November release of* Twilight: Breaking Dawn Part 2 *our index showed positive sentiment toward the movie of 90%. Yet on Saturday, Nov. 24, in the midst of the Thanksgiving holiday weekend, the positive sentiment dipped to 75%. Did that mean consumers were disappointed with the film? Actually, no. We discovered on close examination that many of the people who used words in their Tweets signaling sadness or disappointment were reacting to the emotional moments in the film or to the fact that their beloved series is ending with this installment.*[6]

Sentiment analysis of publicly available data is making inroads into marketing organizations in a variety of industries. For example, bing.com and the Fox Network covered President Barack Obama's State of the Union address, combining the live broadcast with sentiment analytics of opinions and an online voting tool from bing.com. Nearly 12.5 million voters got a chance to express their opinions. These opinions were analyzed and segregated into a number of categories, providing a detailed view of public response to the speech.[7] Gatorade has built a social media command center, where they collect and analyze feedback on their brand.[8]

While this data is extremely useful in detecting trends and patterns, we must be careful in combining this data with other big data. For example, census data provides us with an accurate count of population within a geopolitical area. Can we combine social media data from the geopolitical area to represent the opinion of the entire population? While the numbers of observation points are far higher than a statistically significant sample, it is very much a biased sample representing those who like to express their opinion. It is a highly accurate representation of the subset of population who use social media to express their opinion, but does not represent the rest of the population.

Other than quantification of unstructured text, the posted messages can also be used for discovering underlying patterns and graphs. One such pattern is the influence analysis, which can be measured by the amount of interest a particular blog or tweet generated in its community. There is a variety of ways in which impact of an expressed opinion can be measured and analyzed thereby providing an overall scores for its author. For example, Amazon measures the impact a reviewer has made on product purchases and rank orders these reviewers based on their relative contribution to past purchase decisions.

LOCATION DATA

Next, I would like to discuss location data. We carry our cell phones everywhere, and have now started to use mobile devices to watch movies, browse social media, and make transactions. How can a marketer collect, organize, and analyze location data? I had a chance to work with Jeff Jonas, an IBM Fellow. Jonas was very interested in looking at location data for clues about mobility patterns. In his quest for big data, he stumbled on publicly available mobility data posted by Malte Spitz, a German Green Party politician. Spitz publicly disclosed his mobility information to make everyone aware of privacy issues around location data collected by telcos. He went through a court dispute to collect the data from his wireless phone company, Deutsche Telecom, and made six months of his mobility data publicly available.[9] This data was a gold mine for location researchers. Jonas decided to take the challenge. As a premier expert in the field of customer identity, he was interested in understanding whether or not he could establish the identity of an individual by analyzing mobility patterns.

How was this data collected from Spitz's phone? A cell phone is served by a collection of cell phone towers, and its specific location can be inferred by triangulating its distance from the nearest cell towers. In addition, most smartphones can provide Global Positioning

System (GPS) location information that is more accurate (up to about 20 meters) but can rapidly drain the cell phone battery. In most marketing situations, cell tower location data combined with occasional GPS is good enough. The location data includes longitude and latitude and, if properly stored, can take about 26 bytes of information. If we store 24 hours of location data for 50 million subscribers at the frequency of once a minute, the data stored is about 2 terabytes of information per day. This is the amount of information stored in the location servers at a typical telco. While that is a lot of data, it can be rapidly aggregated to keep only meaningful information.

By itself, longitude-latitude is hard to analyze. I may need to know the granularity of the data and a measure of proximity so that I can infer whether a person is at one location or another. If I have to count the number of people sitting in a building, we need simple measures for location, which can be counted. Fortunately, a number of techniques have emerged for summarizing and counting location. Geohash is one such measure, and is available in the public domain.[10] For a given location using an address or longitude-latitude, the geohash algorithm converts it into a code. The code goes left to right, and each byte further divides the rectangular space represented by the code. A two-byte geohash represents an accuracy of ±630 kilometers, while an eight-byte geohash represents an accuracy of ±19 meters (see table 3.1). The geohash "9x" covers nearly all of Wyoming and northern Colorado, all the way from the eastern boundary facing Nebraska and Kansas to the western boundary facing Utah and Idaho, while "9xj6v0v," an eight-byte geohash represents the corner of Wazee and 20th Street, near Coors Field stadium in downtown Denver. The presence of a person in a specific location for a certain duration is considered a space-time box and can be used to encode the hangout of an individual in a specific business or residential location for a specific time period. By converting longitude-latitude to geohash, I can count how many people were physically present in the vast area covered by Wyoming and Northern Colorado on July 4 at 5 p.m., or do the same analysis

Table 3.1   Geohash accuracy level[11]

geohash length	lat bits	lng bits	lat error	lng error	km error
1	2	3	±23	±23	±2500
2	5	5	± 2.8	± 5.6	±630
3	7	8	± 0.70	± 0.7	±78
4	10	10	± 0.087	± 0.18	±20
5	12	13	± 0.022	± 0.022	±2.4
6	15	15	± 0.0027	± 0.0055	±0.61
7	17	18	±0.00068	±0.00068	±0.076
8	20	20	±0.000085	±0.00017	±0.019

for a street corner in downtown Denver. The first two bytes of these examples are exactly the same—9x as the street corner in downtown Denver is fully contained in the bigger box. However, if we are asked to compute the number people in Colorado, we may need to aggregate a couple of geohashes inside "9x" and "9w," as the state of Colorado is split between those two 2-byte geohashes.

We expected a politician to be fairly mobile and irregular in his mobility patterns. As Jonas began to analyze Spitz's mobility data, he found definite patterns. Spitz moved around a lot, but was still a creature of habit. A small number of hangouts dominated his locations—possibly his home, work, and social meeting places. Jonas used this data and other such studies to establish the identity of an individual based on specific hangouts visited by that individual. This insight is very useful to a prepaid wireless service provider. Most of the subscribers in developing countries are prepaid and often change their phones and subscriber identity module (SIM) cards. However, using hangouts, we can create signatures that accurately represent individuals, giving us an ability to identify them even if they change names, contact, addresses and phone numbers. This insight can be used to identify someone who switches brands regularly and can be used to provide incentives for them to stay with a specific brand.

The discussions with Jonas motivated me to study mobility patterns. I was interested in mobility data for a group of people and to establish insights that can be used for segmentation. In my first job as a market researcher, we used surveys to collect customer data. Surveys can only be administered on a very small (and probably a fairly biased) sample, and are based on recollection of history. We can use cell phone data to collect history, as it happens. I found an app that was able to record my mobility patterns, and after three months of data collection, my pastime was to watch my past mobility patterns through my travels around the world. It was surreal, as if I had a video recording of my past three months of movement. Since it was my own data, I could see the data as well as the inaccuracies. For example, as I drove past a turnpike in Pennsylvania, the tracking was off by a couple of miles, and I could use map data to change the location. As we discovered later, this was a much-needed addition to raw location data, where context information such as street maps can be used to make the data more accurate.

As I work for IBM, I travel extensively, and I expected my mobility patterns to be as wild as water bubbles in a steaming kettle. To my surprise, I showed very definite stable patterns and a small number of hangouts where I spent most of the time. Since my data was coming from my cell phone, it provided me accuracy at the geohash8 level, which is ± 19 meters in the table above. Nearly 15 out of 30 days in a month, I work at home and spend nearly all the time in a single geohash with occasional (and predictable) movements to neighboring geohashes. The other 15 days, I showed mobility patterns that took me from my home geohash to the airport geohash. At that point, the cell phone was turned off and then it was turned on at one of my travel locations. While the distances between these geohashes were large, there were a small number of travel destinations in that three-month period, representing four clients I was working with. In each case, the patterns repeated for each city—representing the hotel, the office location, and the restaurants and bars I was regularly visiting. The analysis of the data also showed me the limitations to using geohash as the mechanism. The IBM office at Armonk is covered by two geohashes, and while I spent most of the time in a single conference room, the occasional trip to the neighboring kitchen to get coffee was depicted as travel to the neighboring geohash 1.22 kilometers away. However, a clever algorithm that predicts velocity of travel allowed me to remove the geohash edge traversal.

I assembled a team of researchers and collected location data from several wireless service providers' location data, on the condition that the data would be anonymized before analyzing, and none of the results would be shared at atomic (subscriber) level. Most certainly, the wireless service providers were interested in the research, but wanted to make sure the demonstrations would not be misused to identify personal information, such as cheating spouses and executives texting while driving (both interesting insights, which can be estimated using location data, although not with 100% certainty). This team of IBM researchers used the data to build a showcase on how the data could be used by marketers.[12] The source data was accurate at the geohash5 level, which means the data was accurate to ± 2.4 kilometers. However, with a little adjustment, Tommy Eunice was able to drive the accuracy down to geohash6 (± 0.61 kilometer). Now we had the data to accurately predict location to the block level. Eunice led the data science work and was rapidly finding clusters of interesting patterns—people who worked from home, buddies who traveled together, or popular lunch places.

Aggregation and clustering are often-used techniques for mining location data. As I stated earlier, geohash coding provides a natural aggregation. A one-byte geohash represents a rectangle bigger than the size of the United States and Mexico (all of the United States and Mexico is represented by the geohash signified by the number "9" and the adjacent geohash "D"). As we add more bytes to the geohash, it divides the bigger rectangle into smaller ones. Once we had the mobility patterns for a large number of subscribers in a city, Eunice was able to aggregate these mobility patterns into two sets of important aggregations. First, he found the popular hangouts by establishing an aggregation of mobility into geohashes at different times of day—early morning, rush-hour commute time, late morning, lunch time, early afternoon, late afternoon, evening commute, dinner time, late night TV time, and so forth. Cell phones and their respective owners congregated at popular hangouts at each of these times, and we could easily spot residential communities, office areas, popular lunch locations, travel congestion spots, and fine dining places. Second, he started to find buddies who traveled together. Two or more cell phones were together in two, three, four, or five locations at the same time. As the number of places visited simultaneously increased in frequency, the data provided us the confidence that these cell phones belonged to people who traveled together and were somehow related to each other. By analyzing the time of day when these cell phones are together, we can predict work, social, or family ties.

Location data can be generated at different levels of accuracy. Typical cell tower data as described above is accurate within 1–2 kilometers. However, wireless subscribers often turn on GPS to find directions on our cell phones. At the expense of a cell phone battery that may get rapidly consumed, the location data captured through GPS is about geohash 8 at an accuracy of 20 meters. Similar accuracies can be achieved when we turn on Wi-Fi in a sports stadium. Wi-Fi location data has one more advantage—in addition to the fact that it does not overdrive the battery consumption, it also improves our ability to connect to the Internet. A public gathering area like a stadium may offer free Wi-Fi to its audience, to ascertain their location data and use it for a variety of operational and marketing purposes. For example, the stadium may offer a visitor advice on which gate to use for entry to the stadium based on current visitor location, seat location, multiple gate locations, and the lines at each gate. There are many interesting marketing opportunities once we have a person with a smartphone located in a stadium who is able to watch the television screen, interact with a little screen, and has a fair amount of interest in buying merchandise located around him/her. By combining more aggregated cell tower data with Wi-Fi data, we can now combine the behavioral characteristics of a shopper (couch potato vs. frequent mall shopper) with the shopping behavior of the shopper (time spent in each aisle or combinations of aisles visited). A savvy data scientist can also use clustering algorithms to establish micro-segments by finding individuals who follow similar mobility patterns. Some of the micro-segments are based on people traveling to similar locations. However, more complex micro-segments are based on mobility patterns to diverse locations. For example, a statistical program can find active weekend golfers who wake up early on the weekend and show up at the golf course for a Saturday morning game. These golfers may be showing up at different golf courses around the globe, but share the Saturday morning mobility pattern. This micro-segment is of enormous interest to golf companies, golf resorts, and the leisure travel industry.

How about combining social media and location data? If there is a way to correlate social media data to mobility data, it can provide marketers with a valuable cross-correlation of customer profiles. To examine this, a team at IBM's Global Solution Center collected two months of Twitter data and performed a series of unstructured analytics.[13] They also had access to the mobility data described above. A marketer might like to find people who show specific travel patterns and who tweet about sports to offer them sports memorabilia at their next visit to the stadium. Finding people who like sports is not as easy, however, as tweets come with different words that describe sports. Once a data scientist has found a baseball fan, the next interesting challenge is to align it to the mobility data. Unless the correlation is done with full disclosure to the customer, this task may not be appreciated by the customer. In many cases, the stadium may have complimentary Wi-Fi and may trade free Internet access for wireless information and a Twitter handle, possibly offering a sports statistics app as a promotion. Now we have access to all the Twitter information from this Twitter handle to make guesses about the person behind the Twitter handle. We can also use the mobility data to find additional micro-segments. We can find their buddies and start offering products and promotions that appeal to the consumer or to his/her social circle. If someone tweets "Enjoying a Rockies game with my hubby" and is a frequent visitor to Rockies games who works in downtown Denver, the marketer can easily infer: "married," "woman," "enjoys baseball," "frequent Rockies game visitor," and "daily grinder," and offer specific promotions that may appeal to this person.

I earlier discussed census data, which is the most comprehensive big data source. The mobility data discussed here is still a sample, as it represents only cell phone users and only those who subscribe to a single wireless provider, unless we start combining data across wireless providers. However, cell phone data provides a marketer with an observed count of "work at home" in a geographical area. What if we could pick a statistically significant sample of individuals from the same geographical area and ask them if they work from home? The reported information would not be as accurate as observed data because, depending on the exact phrasing of the question and how the respondent interprets it, the data may not be as accurate as observed data. However, the observed data from a wireless device is a good representation of the wireless phone users for the specific wireless provider, but not necessarily the rest of the population. That data would fail to represent my 90-year-old retiree dad, who does not carry a phone and stays at home most of the time.

PRODUCT USAGE DATA

With all these powerful contributions to data-driven customer profiles, let me take the next source of big data—product usage. As products become digital, they contribute usage data, which can be analyzed and used for customer modeling. Let me take you through a couple of examples to describe the extent of big data available from product usage and how it can be analyzed and potentially combined with social media and location data.

For a number of decades, television producers relied on a control sample of audience viewing habits to gauge the popularity of their television shows. This data was collected using, first, extensive surveys in the early days of television programming and then, later, special devices placed on a sample of television sets by companies such as Nielsen. With the advancement in the cable set-top box (STB) and the digital network supporting the cable and satellite industries, cable operators can now collect channel-surfing data from all the STBs capable of providing this information. As a result, the size of data collected has grown considerably, providing marketers with finer insights not previously available. This information is valuable because it can be used to correlate channel surfing with a number of micro-segmentation variables.

Television STB data is available at the household level, while mobile device content data provides content viewing by individuals. Consumers are beginning to watch content on both platforms, and sometimes they even use both at the same time for complimentary information access. This data is collected from the devices, cleaned up, and correlated with programming data to ascertain the timing of customer behavior. If I use a two-way STB to watch television, the supplier has instant access to my channel-surfing behavior. Did I change the channel when the advertisement started? Did I turn the volume up or down when the commercial started to play? What if the consumer started to watch the television, but left it on for the day while going to work (maybe while turning off the television, but not the STB). A fair amount of cleanup is needed before this data can be analyzed. STBs are geographically located. If we know the television viewing habits of a community of people, that information can be utilized for beaming specific messages to that community. Aggregation and correlation can be used to analyze STB location data, combined with STB usage data.

As wireless devices get smarter, agents installed on these phones collect device usage information to analyze this data for device or network quality improvement. Sometime ago, my iPhone 4S was showing erratic behavior. It would start heating up all of a sudden and drain its entire battery in a short time, leaving me without a working phone.

I made an appointment with the genius bar at Apple and showed up with the phone. The genius bar representative connected my phone to his laptop and could see my battery temperature history. He concluded that a bad app was causing this erratic behavior and that the best remedy was to remove all the apps, reset the phone, and reinstall a new version of the apps one by one. CarrierIQ offers similar services for Android phones and has been providing device analytics to wireless service providers. By most industry measures, the numbers of smartphones being returned by customers in the first year are running at an unacceptable and unsustainable rate. Perhaps unsurprisingly, the rates appear to vary dramatically between handsets, but appear to be averaging 15 percent–20 percent. What is surprising is that over 40 percent of these devices turn out, upon further investigation, to have nothing wrong with them. This insight was echoed in the recent IWPC Mobile Field Returns Survey (September 2011), in which a selection of US and European operators were asked to report on the volume and type of field returns. Again, no-fault-found ranked in the >30 percent category from most operators.[14]

Once this data is collected, can it provide any value to sales and marketing? Many consumers live in houses or apartments with poor wireless coverage. We often tell our callers to call back on a landline so that we can reasonably converse without poor wireless quality or dropped calls. Network equipment providers and wireless operators have worked together to provide network devices, which can be attached to a broadband Internet connection to provide wireless signals.[15] Since these devices use broadband access, the usage is no longer counted as minutes connected, and the device shows a whopping five-bar network connectivity. How do I identify a micro-segment—miserable subscribers who get poor network coverage near their house? This is a good example of utilizing mobility pattern information to identify residential accommodation for a subscriber and device data to establish poor-quality network coverage. Once joined, the intersection of the two is a target list to market "home connects"!

Regulators have asked most telcos and cable operators to store call detail records and associated usage data. For a 100-million-subscriber telco, the Call Detail Records (CDRs) could easily exceed 5 billion records a day. Telecom operators have been collecting CDRs for a long time. As of 2010, AT&T had 193 trillion CDRs in its database. As phones became more sophisticated and consumers started to use the phone for activities other than calling, the CDRs started to include other forms of communications, and everyone started to use the term xDR, where x is a variable that takes many meanings, depending on whether it represents calls, text, data, or video. Big analytics has lately started to provide telcos with sophisticated capabilities to analyze this data and find useful nuggets of information about their customers.

Social groups can be inferred from any type of communication—emails, SMS texts, calls, Facebook friendships, and so on. It is interesting to see strong statistics associated with leaders' influence on their social groups. One such analysis involves discovering group dynamics. Communication across individuals can give insights into formal and informal groups. In some situations, these groups have formed among coworkers, and could be very formal with organizational hierarchies and matrices. In other situations, the communications may be due to informal social groups formed via families and friends. In any group, there are leaders who keep these groups together and followers who are influenced by those leaders. There may be ambassadors who may belong to one group but represent them in another, where they have loose ties. There are many ways to discover these groups by using big data. Telcos are rapidly discovering that they have a gold mine of big data in the form of xDRs with social/work group information, which can be used for marketing purposes.

Social group leaders typically have a set of social group followers. If these groups are communicating with each other, it is a possibility that the brand choices made by the leaders will influence the subsequent brand switching among the followers. Let me use an example to illustrate this behavior. I am a member of an investment club, which regularly meets to discuss investment decisions. We have pooled together a small fund, and we make decisions about buying and selling stocks using the pool of funds. In addition, each of us has our individual investments. Our group decisions often influence our individual decisions, especially in dealing with the investment brokers and tools. Collectively, we use five brokers for individual investments across the group. As the group leaders start switching from one set of brokers to another, others start to follow. The group communicates regularly with each other using cell phones, with a significant number of calls and texts. Can we analyze the xDR data to predict brand switching for investment brokers? Once a leader switches a brand, it increases the likelihood for the social group members to churn as well. Who are these leaders? Can we identify them? How can we direct our marketing to these leaders?

In any communication, the leaders are always the center of the hub (see figure 3.1). They are often connected to a larger number of "followers," some of whom could also be leaders. In the figure, the leaders have many more communication arrows either originating from or terminating at them compared with the others.

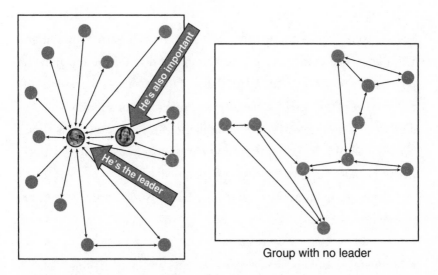

Figure 3.1   Leaders in a communications network

How do we identify the leaders? IBM Research conducted a series of experiments with telcos. CDRs, which carry information about person A calling person B, were analyzed. By synthesizing call information and abstracting communications networks, we discovered webs of communications across individuals. We also used the customer churn information to correlate churn among leaders to subsequent churn among followers. Here are some of the highlights from one of the experiments that I helped conduct:[16]

- Leaders were 1.2 times more likely to churn compared with non-leaders.
- There were two types of leaders: disseminating leaders who were connected to their group through outgoing calls, and authority leaders who were connected through a larger proportion of incoming calls.
- When a disseminating leader churned, additional churns were 28.5 times more likely. When an authority leader churned, additional churns were 19.9 times more likely.
- Typically, there was a very limited time between leaders' churn and the followers' churn.

Social group data is increasingly available from a variety of sources. This data is becoming an important source of data for the social media products, in addition to the telcos. Each of these sources represents different types of groups. The overall social dynamics of a group is a combination of observations across a number of sources. A set of Facebook friends may not be communicating via phone, but could be actively sharing product evaluations using social media sites.

There are many ways to utilize social networks to influence purchase and reuse:

- *Studying consumer experience*—A fair amount of this data is unstructured. By analyzing the text for sentiments, intensity, readership, related blogs, referrals, and other information, we can organize the data into positive and negative influences and their impact on the customer base.

- *Organizing customer experience*—We can provide reviews to a prospective buyer, so he/she can gauge how others evaluated the product.
- *Influencing social networks*—We can provide marketing material, product changes, company directions, and celebrity endorsements to social networks so that social media may influence and enhance the buzz.
- *Feedback to products, operations, or marketing*—By using information generated by social media, we can rapidly make changes in the product mix and marketing to improve the offering to customers.

SHOPPING DATA

Sales and marketing got their biggest boost in instrumentation from the Internet-driven automation over the past ten years. Browsing, shopping, ordering, and providing customer service on the Web has not only provided tremendous control to end users but also created an enormous flood of information to the marketing, product, and sales organization in understanding buyer behavior by analyzing usage data. Each sequence of web clicks can be collected, collated, and analyzed for customer delight, puzzlement, dysphoria, or outright defection, and the sequence leading to this decision.

Self-service has crept in through a variety of means: IVRs, kiosks, handheld devices, and many others. These electronic events act like a gigantic pool of time-and-motion studies. We have data available on how many steps a customer took, how many products he/she compared, and what he/she focused on: price, features, brand comparisons, recommendations, defects, and so on. Suppliers have gained enormous amounts of data from self-service, electronic leashes connected to products, and the use of IT. If I use the Internet to shop for a product, my click stream can be analyzed and used to study shopping behavior.

How many products did I look at? What did I view in each product? Was it the product description or the price? This enriched set of data allows us to analyze customer experience in the minutest detail.

A number of companies are collecting detailed click-stream data to understand consumer behavior. The motivation for collecting this data came from understanding how customers were using websites, and whether or not there were glitches in the web design that prevented customers from using the sites for the intended function. Whether direct-to-consumer or business-to-business (B2B), it is no secret the online channel is a critical component of business today. Yet analysis of website usage patters identifies users who regularly struggle to complete transactions online and as a result abandon their online transactions midstream. Tealeaf is an example of a product that captures the qualitative details of each interaction.

While clickstream data has provided web designers useful tools to design better websites, it has provided a great source of big data to marketers. As products and promotions are introduced in the marketplace, clickstream data provides marketers with specific information regarding the details of information access. How many customers visited a specific page? What was the sequence of clicking before and after clicking a specific page? How much time was spent on a specific page? If there were actions available based on the information shared, did the audience follow the link or not?

Clickstream data can be correlated with other data. If the clicks were made on a mobile platform, it would be interesting to note the geo-hash location of the mobile device at the time of the click, and whether or not someone else was also present among the consumer's buddies. Often consumers are clicking on websites while at the same time watching content on television in a direct response to a commercial and then sharing among social groups. I remember conducting and using a commercial day-after-recall survey mastered by Procter & Gamble (P&G) and Burke Marketing[17] in the 1970s and 1980s. The correlations we can make today are a significant enhancement over day-after-recall as we can track consumers with clickstream data in conjunction with the airing of a commercial. In addition, with a series of profile labeling and drill downs, a marketer can analyze the impact by micro-segment and establish an accurate persuasion measure for a commercial using a large proportion of the population.

Shopping is becoming more sophisticated. I recently visited Home Depot and Best Buy to shop for kitchen appliances. By downloading their respective apps on my smartphone, I could collect a fair amount of additional data on the appliance while standing in front of it. I was using the Wi-Fi service supplied by each of these stores, which means they can effectively track my movement within the store and correlate the movement and the use of the app to accurately predict my shopping list. The smartphone identification data, in conjunction with the app use, can allow marketers to correlate shopping with specific customers. If they have access to the audience data from their cable operator, and to browser data from a clickstream supplier, they can correlate a fair amount of consumer information, making a judgment about the level of information that has already been shown to this consumer and whether there are significant pieces of information that have not yet been presented.

CONVERSATION DATA

Customer touchpoints provide valuable data to customers shopping for products. Is this a data source for marketing? Would it be possible for us to capture call center conversations, web chats, email trails, and so forth to gauge customer interest? Can we also use this data to understand differences across customer cross sections?

Marketers are increasingly using customer touchpoints to communicate with customers, and a fair amount of this data is already being captured for further analysis. A cable provider recently asked me to analyze their call center data to ascertain customer intention. In their environment, the call center agents were codifying customer intent using 150+ codes at the end of the call. As suspected, the reported information was not necessarily accurate. Most of the call center agents used fewer than ten codes to represent most of their conversations. It was not clear whether customers were calling only for those reasons or whether the call center agents could memorize only a small number of codes and were repeatedly using those codes to inaccurately represent the real conversation.

However, a mechanical means of data categorization and mining can be used to verify and autocorrect these observations. Call center conversations can be converted into text, and then the text can be categorized into a code. In addition, the customer voice analysis can detect anger or appreciation, leading to not one but possibly many quantifications of the call center information. Web chat and email can also be analyzed, and although they lose the verbal emotions, they can be used for written emotions, for example, use of adjectives and adverbs to codify emphasis and emotions.

The most powerful analysis from conversation data is in its use to identify gaps between intention and action. If someone is interested in purchasing a product, but does not end up buying in a specific contact point, it represents an unfulfilled demand that can be further addressed via a campaign.

PURCHASE DATA

I am working with a set of mobile wallet technology suppliers, and they have given me some insights on the rich data we can collect from purchases. There are several sources of this data. Credit card providers carry information about credit card transactions. Their data contains merchant, transaction location, and transaction amount data tracked by consumer. Mobile wallets carry similar data, and sometimes have additional money movement data if the wallets are being used for lending money, especially in the growth markets where currency and banks are getting replaced by mobile wallets.

With online transactions, we are beginning to see technologies that can help consumers and marketers organize the purchase data. I use

Slice (www.slice.com) to keep track of my online purchases. Slice scans my email for any online purchases and extracts relevant information, so I can track shipments, order numbers, purchase dates, and so on. I do a fair amount of business as well as leisure travel and often make advanced reservations for airlines and hotels. Slice provided me with a convenient app on my iPhone to track all of these reservations. With a couple of clicks on this app, I can get access to valuable data about my confirmation number, date of reservation, and the hotel address and phone number. Slice extracts the relevant ordering information and keeps it organized for me for easy access to this data.

Slice also lets me "slice and dice" the orders. That is, it analyzes my purchases against a set of categories to report the number of items and money spent in each category. Figure 3.2 shows Slice's category analysis: Travel & Entertainment, Music, Electronics & Accessories, and so on. In doing so, Slice is doing rigorous unstructured analytics and user interaction to identify what is considered "Movies & TV" and how that is different from "Music."

The classic product categories originated from the Yellow Pages. We remember the classic Yellow Pages books that we received yearly and that are nowadays being incorporated into online Yellow Pages and other shopping and ordering tools. However, categories are typically tree structured, where each node is a subclass of the node above and can be further subclassified into further specialized nodes. For example, a scooter is a subclass of a two-wheeler, while an electric scooter is a subclass of a scooter. A node can be a subclass of more than one entity. A subclass shares the attributes of its superclass. Therefore, both scooters and electric scooters should have two wheels. While the classic product catalogs were static and were managed by administrators without organized feedback, the unstructured analytics provide the ability to make a dynamic hierarchy, which can be adjusted based on usage and search criteria.

As companies like Intuit and Slice deal with their users, they provide a categorization of a transaction based on their collective understanding

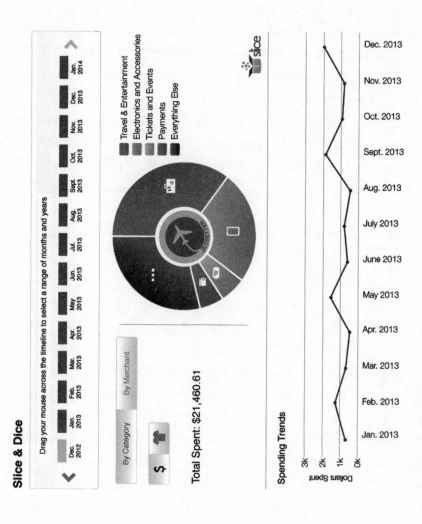

Figure 3.2   Slice and dice of my purchase data

of the product catalog. However, they provide the consumer with the ability to change the classification. The analytics of reclassification allow these algorithms to constantly adjust their classification based on real data from their customers.

The data that Slice is collecting provides significant value to marketers. Here is a *Washington Post* analysis of iPhone sales from a recent launch of iPhone 5:

> *The sales figures outstrip analysts' expectations for the opening weekend. Estimates for the phones' sales ranged from 5 million—the number of iPhone 5 models Apple sold on that model's opening weekend—to 8 million. Apple didn't specify which iPhone colors were the big favorites with consumers, though more anecdotal reports suggest that the gold iPhone 5s was far more popular than the silver or "space gray" models. The gold version was the first to run out on Apple's Web site, and several customers reported that they were having trouble finding the phone in stores if they weren't at the very front of the line. As for the more colorful, plastic-backed iPhone 5c, at least one firm estimates that nearly half of its customers favored a blue or pink phone. Slice, an online firm that helps users track their online purchases, reported that 28 percent of pre-orders it tracked for the iPhone 5c were for blue phones, 20 percent for pink phones. What was the least popular color for the iPhone 5c, according to Slice's data? Yellow, which accounted for 10 percent of the orders.*[18]

Grocery stores have been equally busy developing their understanding of customers. Most of them offer frequent shopper cards that can be used by the grocers to track purchase habits as well as used by shoppers to redeem discounts and other useful campaigns. With identifying information collected from the customer, this shopper card can be correlated with a name and an address. Retailers toyed with the idea of providing shopping gadgets to shoppers and eventually realized that creating a smartphone app to run on an existing device would be easier than engineering a new device. Shoppers may activate a mobile app as soon as they enter a retail store. The app starts to collect GPS-level accurate location information about the shopper and lets him/her check in grocery items on the smartphone. At the checkout counter, the shopper connects the smartphone to the point-of-sale (PoS) device, and the grocery bill is automatically paid by the credit card associated with the app. As the person walks through the grocery store and checks in grocery items using a smartphone, a campaign management system starts downloading mobile coupons based on customer profile, past grocery purchases, and currently active promotions.

PROPOSITION

In 1980, I was conducting day-after-recall research for advertisements aired on television. A typical day-after-recall requires finding a statistically significant sample of television viewers who watched the programming the previous day and asking them many questions about what they remember seeing. The advertisement was considered successful if a large proportion of those who saw the advertisement remembered it as well as its marketing message. As I started to collect the data, I used census information to design our survey collection, providing adequate coverage of various income groups in the city of Mumbai, and trained a number of my interviewers on the survey questionnaire. To understand the data collection accuracy, I followed a couple of interviewers throughout the day all over the city. Dogs and security guards often chased us, and many prospective respondents slammed the doors before we could ask any qualifying questions. At one house, the main decision-maker did not have the time to answers, so she directed me to her teenage daughter who was eager to answer the questions, but was not the primary decision-maker. My field interviewer and I argued forever about the validity of that observation. He told me the teenage daughter met the criteria specified in the interview, and so the interview was valid. I kept thinking about their next trip to the grocery store and the role the daughter was likely to play in using the advertisement to decide on the product purchase. As the day progressed, I started to get a realistic view of the "statistically-significant perfectly-random sample." Irrespective of how hard we tried, the sample remained biased toward those who were eager to respond and were easily accessible to us.

That was 33 years ago. As I recollect the experience, it really feels like another century! Today, I have some of that data available from the STB of all digital cable subscribers in the city, so instead of chasing 100 subscribers, I can be looking at data from 5 million subscribers. I can use the data to identify subscribers who saw an advertisement and determine whether they reached for the remote halfway through the advertisement to switch the channel or reduce the volume, and by analyzing the social media messages, I can seek their sentiments about the commercial. Yes, that is still a biased sample, but it represents a much bigger sample size. Depending on the geography, the STB data may represent a biased majority, and the social media messages only belong to those who are eager to respond, like the teenage daughter of the busy housewife I interviewed 33 years ago. The difference is we now have a lot more observations, not just reported samples of data. Also, we may find an overlapping set of data. Each big data source brings its own biases, but truly represents an individual. Despite these biases, we may be able to map a set of customers almost perfectly—where they dine, what television programs they watch, how far they commute to work, when they take coffee breaks, which brands they prefer, and how they respond to different campaigns.

So, how does this change marketing? For decades, statisticians build processes that worked well on random samples of real data. We now have real data. There are no more samples. Also, if we are able to build a relationship with a customer, we can track that customer through different stages of purchases.

Marketing is about making customers aware of the offerings, supporting the buying process via a variety of persuasions culminating in a purchase, and then using this affinity to sell the next product, expand to his/her circle of friends, or design a new product based on those ideas. This chapter established the first proposition, that marketers have a lot of observations they can use for anything they would like to do. It also provided a new task for statisticians, to work on systematic biases and remove their bad effects. Now that we have found a new frontier where there are no more small samples and where marketers have access to enormous observations about each customer, do we continue to broadcast messages to our customers? In the next chapter, I will explore the actions a marketer can take and how big data radically changes how marketers interact with their customers.

4

FROM BROADCAST TO COLLABORATION

INTRODUCTION

In the last chapter, I built the first proposition, showing how big data gave us a lot more observations about customers. This rapid rise in data has also been coupled with an equally rapid advancement of advanced analytics and automation to drive marketing decisions. The market leaders foresaw the availability of big data and started to build gigantic receptacles to contain and control the big data tsunami to their market advantage. A large number of consumption options have fueled the need for real-time and intelligent decisions, which must be automatically generated and fed into the consumption engines. A number of these advancements discussed here are very disruptive to the market, as they are driving significant automation and disintermediation of the middlemen, who processed and massaged the data. They are tearing apart marketing as we knew it in the past and replacing it with a new set of actions.

I continue to receive nearly five to ten promotions from credit card companies per week in the mail. Each of these is an invitation for a new credit card with lucrative sign-in bonus miles or other goodies. As I toss those offers unopened in the trash, I sometimes try to estimate the cost of these campaigns and imagine their intake reports. Is anyone with my profile using these promotions? How can the credit card companies improve their yield? It so happens, like many other empty nesters, I have no need to shop for credit cards. Each time I switch credit cards, my credit rating companies reduce my credit score, and I need the best scores to keep my mortgage rates low. Last but not the least, I have not used any of the offers provided to me via mail, ever! Do they not have a way of tracking a completely disinterested prospect? Maybe I am not the best target for these promotions? Can they track my response and improve their campaign yield?

Marketing organizations have traditionally broadcasted their campaigns to customers based on their analysis and understanding of segments. As long as a reasonable proportion of customers from the target population was reacting favorably to the offers, marketers kept investing in the rest of us, filling mailboxes and wastebaskets with promotions that were never read or acted upon by consumers.

The early evolution was in the use of analytics for segmentation. The original segmentations were demographic in nature and used hard consumer data, such as geography, age, gender, and ethnic characteristics to establish market segmentations. Marketers soon realized that behavioral traits were also important parameters in segmenting customers.

As our understanding grew, we saw more emphasis on micro-segments—specific niche markets based on analytics-driven parameters. For example, marketers started to differentiate innovators and early adopters from late adopters based on their willingness to purchase new electronic gadgets. Customer experience data lets us characterize innovators who were eager to share experiences early on and might be more tolerant of product defects.

In the mid-1990s, with automation in customer touchpoints and use of the Internet for customer self-service, marketing became more focused on personalization and 1:1 marketing. As Martha Rogers and Don Peppers point out in their book *The One to One Future*, "The basis for 1:1 marketing is share of customer, not just market share. Instead of selling as many products as possible over the next sales period to whomever will buy them, the goal of the 1:1 marketer is to sell one customer at a time as many products as possible over the lifetime of that customer's patronage. Mass marketers develop a product and try to find customers for that product. But 1:1 marketers develop a customer and try to find products for that customer."[1]

Social media created the next wave of changes, which significantly improved consumer power through collaboration. Consumers started to collaborate—first for social reasons, but then to compare notes on marketers. Yelp created the crowdsourced rating system in which consumers can share their experiences. Amazon started to pay attention to the most influential reviewers who changed the opinions of others. Facebook started as a social media site first, but then gradually opened its vast customer base to marketers.

Fast-forward to today. Campaign delivery capabilities have improved significantly. With online purchasing gaining critical mass, there is a need and an opportunity to use marketing at electronic point of sales at silicon speed—in milliseconds. These marketing capabilities are fueling three significant improvements in marketers' ability to influence customers and vice versa. First, marketers are able to use predictive modeling and social media to find the customers best suited for their campaigns. Unlike reporting systems of the past, sophisticated predictive models mine the data to target specific customers and sharpen the messaging to them. Social media offer ways to organize customers and offer customer groupings to marketers. The second improvement is in our ability to develop marketing actions at high speed and use them to influence targeted customers in real-time or on-demand. These actions can be delivered to specific customers or micro-segments. Third, is our ability to collect customer reaction to a focused campaign and fine-tune the campaign based on the reaction. It allows the consumer to influence marketers. By providing a "thumbs up" to an advertising message, consumers can communicate back to the marketer, their interest in the subject area. These three improvements—micro-segmentation, focused messaging, and customer feedback—provide the necessary pillars for collaboration between marketing and customers and drive influence in marketing campaigns.

Marketers have used a variety of ways to influence customers. In this chapter, I will focus on a number of these marketing capabilities and show how they are bringing dramatic changes to marketing capabilities in order to reach and influence customers. I will cover their impact on product design, advertising, promotions, and pricing.

PERSONALIZED CUSTOMER / PRODUCT RESEARCH

I invested $100 in campaign contributions to Barack Obama's campaign before the 2012 presidential election. The personalized interactions have been among the best I have had with any political organization. Here is an example of something no other politician has done for me. I received a letter from Michelle Obama. First, the letter mentioned the progress the Obama administration has made in immigration reform and climate change—two issues I passionately care about. Next, the letter asked me, "What is the number-one issue you care about?" The letter thanked me for sharing my opinions in the past and encouraged me to keep the communication going. As I have responded with my top issues, the campaign has continued to refine its messages and keep track of my issues, keeping the messages personalized and tracking the president's actions in response to my concerns.

What we are seeing here is a fine example of consumer research at a personalized level. Customers interact with a number of companies that supply goods and services to us. How often are our top issues tracked by them? I was grumpy about the slow speed of my Internet connection and saw an advertisement from my telecom provider offering me five times the broadband speed for an Internet connection. I called them to ask if I could subscribe to the faster connection. The call center agent told me the service was not available in my area. He casually told me to call back again in the future to check whether or not the broadband service would be available. I asked him whether I should call every five days until the service is available. He was not sure, but said calling once a month was not a bad idea. Each such call would cost this company about $20, and they would not collect any information from their customers to find how many in a given area were calling for higher bandwidth. If I call them five times about this upgraded line, they will have incurred enough call center expenses to forgo any profits for selling the upgrade to me for the first year. They were ignoring simple indicators of customer demand, which they could have used to fine-tune their bandwidth demand and product availability across geographies. The only tracking they required was a list of all customers who had expressed the need for higher bandwidth. As soon as the bandwidth was available, they could connect with customers and offer them an upgraded product. In addition, now they would have demand information for their new upgraded service, which could be used by their network engineering group in deciding where to build the network infrastructure for the upgraded service.

I can recollect a similar story about my favorite airline. I was seeking a seat on a particular flight, and the response I received was identical to the one I received in the example above—"Please call back once a month, and you will hopefully find the seat you are seeking." An airline can be very sophisticated in their service, especially as they deal with their elite customers. They can easily track what their best customers would like to do and offer them customized packages based on customer needs. All of these are examples of customized product design based on customer requirements. Marketers often cater sales and order processing to the products they have rather than to the needs their customers have. In a micro-segmented, or personalized, marketing situation, they can build a set of customer profiles and offer customers products based on what they need or on what similar customers are buying. As you may have already discovered, this is what Amazon already does. Every week, I receive a list of books that Amazon recommends to me to buy, which is based on my past purchases, new books released, and

the purchase behavior of similar customers. Amazon's recommendation engine mines through past purchases and book classifications to
build the recommendations.

Customer usage is the best source for customer and product
research. As marketers offer complex products with many features, and
often find customers not using many of the features, a product can be
rationalized. That is, by analyzing product usage, a product manager
may drop product components, features, or accessories that no one is
buying. In a study with a telecom provider, I found that 98 percent of
their customers purchased 197 product components or features out
of nearly 20,000 offered to their customers. As the product managers
insisted that they were dealing with a fat tail, I extended the analysis to
99 percent of the customers and found only 500 product components
or features in use. Their call centers were training their sales personnel for six weeks, most of the time teaching them how to enter 20,000
product codes in their order-processing systems. While the rest of the
product components did not offer any exceptional margins or market
advantage, they were available for an occasional buyer. As the company
began to introduce a simplified product line through their sales channels and relegated the "once in a blue moon" product components to
a specialized sales process, their sales training time was dramatically
reduced from six weeks to two days.

Products are often designed to comprehensively cover all customers. With any office software tool, like Microsoft Word, which I am
using to write this book, most users employ a very small number of features, and a couple of power users employ a specialized set of features.
Could we custom design products based on customer needs and offer
additional components as a customer requires those new capabilities?
Product usage analysis can help us map product features to customer
groups and simplify offerings targeted to those groups.

Product automation provides an enormous opportunity to measure customer experience. Today's sophisticated consumers take photos
digitally and then post them on Facebook, providing an opportunity for

face recognition. They listen to songs on Pandora, creating an opportunity to measure what they like or dislike, or how often they skip a song after listening to the part of it that they like the most. They read books electronically online or on our favorite handheld devices, giving publishers an opportunity to understand what they read, how many times they read it, and which parts they look at. They watch television using a two-way set-top box that can record each channel click and correlate it to analyze whether the channel was switched right before, during, or after a commercial break. Even mechanical products such as automobiles are increasing electronic interactions. These customers make all of our ordering transactions electronically, giving third parties the opportunity to analyze their spending habits by month, by season, by ZIP+4, and by tens of thousands of micro-segments. Usage data can be synthesized to study the quality of customer experience, and can be mined for component defects, successes, or extensions. Analysts can identify product changes using this data. For example, in a wireless company, analysts isolated problems in the use of cell phones to a defective device antenna by analyzing call quality and comparing it across devices.

Products can be test-marketed and changed based on feedback. They can also be customized and personalized for every consumer or micro-segment based on consumers' needs. Analytics plays a major role in customizing, personalizing, and changing products according to customer feedback. Product engineering combines a set of independent components into a product in response to a customer need. Component quality impacts overall product performance. Can product managers use analytics to isolate poorly performing components and replace them with good ones? In addition, can they simplify the overall product by removing components that are rarely used and offer no real value to the customer? A lot of product engineering analytics using customer experience data can lead to the building of simplified products that best meet customer requirements. The solution requires a data-driven mapping of customer needs and product usage to product components. The mapping can be utilized by product marketing to offer product packages,

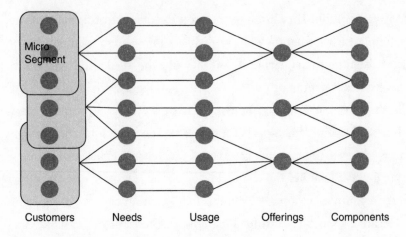

Figure 4.1   Mapping of Customers and Products

bundles, and customizations to specific micro-segments. The mapping can also be used by product engineering to change product components based on customer needs and product usage. Figure 4.1 depicts a data-driven mapping, which can be deduced from the types of observations described in chapter 3 and used for marketing segmentation, product marketing, and product engineering.

The first set of dots represents customers. These customers exhibit certain needs, depicted by the second set of dots. Customers may use a variety of means to communicate their needs, such as using social media to tell friends they are interested in purchasing a product, or by searching for product-specific information on the web. The links between customers and needs can be derived by statistical analysis of observations and can be depicted as strengths in the lines connecting customers to needs. A need is associated with a usage, which represents how the need is fulfilled. Thus need to work at home (a need) may be related to use of high bandwidth (a usage), as these users consume high bandwidth as they share presentations via email and use corporate applications. The usage can be linked to product offerings from a marketer, typically in the form of a grouping of components, which are sold together. The marketer may group these customers based on needs and usage, and may develop specific offerings made up of product components to respond to these customers. In my earlier example, I described my need for a higher bandwidth Internet connection to my telecom provider. In addition, I may be showing higher usage of bandwidth in the daytime on weekdays, and average use outside of business hours. A telecom provider may group such customers into "daytime work at home" (a customer grouping or micro-segment) develop an offering for "higher bandwidth during weekday," using product components available from the engineering organization. The engineering organization may employ these offerings to create engineering components that offer different bandwidth by time of day to different customers, based on their product subscriptions, by combining components—"bandwidth" and "higher bandwidth network policy" (see figure 4.2). Needless to say, this offering may provide additional revenue to the telecom provider, make use of idle bandwidth during the daytime, when the rest of the neighbors are "daily grinders" and commute to a work location, and lead to a higher loyalty rate among people who work from home. Product marketers can discover many such micro-segments by analyzing the data. They can also offer products based on these segments, focus their campaigns on a targeted set of customers who are exhibiting specific behaviors, and observe the intake for those products in the targeted segments. The links shown in this figure represent observations and data sets and can be derived by mining data observed from customers. Once the model has been identified, it can be used for targeted

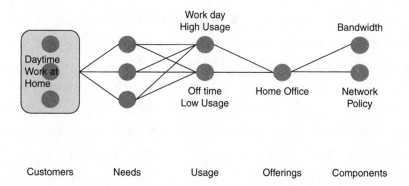

Figure 4.2   Daytime Work at Home Micro-segment

campaigns to specific customers, or for designing new offering, by combining product components.

Intelligent segmentation and campaign management systems based on these approaches have resulted in significant uptake in revenues and customer satisfaction. The campaigns developed using these techniques were far more focused, and also far more successful, in comparison to broader campaigns. Prior to product automation, data collection was very difficult. However, recent advances in product and touchpoint automation have given rise to observation data, which can be collected and analyzed without significant investment.

The same approach can also be used for product rationalization. Marketers offer a large number of products and components to their customers. Usage observations are key to identifying and isolating sporadically used product components and features. The analysis can also be focused on badly designed components, which need to be either redesigned or removed from the product mix. Product managers may use profitability analysis, along with competitive intelligence, to decide which features are not adding value to product selling or usage, and drop components that offer no leverage to the product mix.

To conduct this analysis and predictive modeling, we need a good understanding of the components used and how they participate in the customer experience. Once a good amount of data is collected, the model can be used to isolate underutilized or badly performing components by isolating the observations from customer experience and tracing them to the component. Complex products such as automobiles, telecommunications networks, and engineering goods benefit from this type of analytics around product engineering.

The first level of analysis is in identifying a product portfolio mix and its success with customers. For example, if a marketer has a large number of products, they can be aligned to customer segments and their usage. We may find a number of products that were purchased and hardly used, leading to their discontinuation in six months, while other products were heavily used and sparingly discontinued.

Once we have identified less-used products, the next analysis question is whether we can isolate the cause of customer disinterest. By analyzing usage patterns, we can differentiate between successful products and unsuccessful ones. Were the unsuccessful ones never launched? Did many users get stuck with the initial security screen? Maybe the identification process was too cumbersome. How many users could use the product to perform basic functions offered by the product? What were the highest frequency functions?

The next level of analysis is to understand component failures. How many times did the product fail to perform? Where were the failures most likely? What led to the failure? What did the user do after the failure? Can we isolate the component, replace it, and repair the product online?

These analysis capabilities can now be combined with product changes to create a sophisticated test-marketing framework. We can make changes to the product, try the modified product on a test market, observe the impact, and, after repeated adjustments, offer the altered product to the marketplace.

Let me illustrate how big data is shaping improved product engineering and operations at content providers—cable companies and telecom providers that are providing regular cable channels, over-the-top content, Internet Protocol television (IPTV), on-demand, and so on. For many decades, the cable infrastructure was essentially a lot of fat pipes connected to a cable company's content hub where cable employees ran around on roller skates to change contents as requested by the consumers. All this changed with the DOCSIS 3.0 (Data Over Cable Service Interface Specification) standard that started to offer digital content over high-bandwidth digital pipes. In the meantime, telecom providers started to offer IPTV. Also, Netflix, Google, and Apple began offering content on the Web, which could be displayed on the regular television. Interactive television has radically changed the game for the entire content industry. The content is no longer broadcast to a set of homogeneous channels. Consumers have the ability to customize their content, fortunately, under the minute scrutiny of the content provider. Cable operators and telecom providers collect enormous amounts of data about the network, including network transport information coming from the routers and the switches, as well as usage information, which are recorded each time we watch content on a screen. For larger cable and telecom providers, the usage statistics are not only high volume (in billions of transactions a day) but also require low-latency analytics for a number of applications. This data is quite valuable for recommending new content, placing advertisements during the viewing, and designing new programming by content providers.

Netflix rose as a viable competitor to cable and telecom providers. Starting from a DVD mail-order business, Netflix has rapidly grown into an online content provider with on-demand customized content it offers to its subscribers through a monthly subscription program. As customers use Netflix services to watch content, their usage data is meticulously collected, sorted, stored, and used for analytics to provide content recommendations. The Netflix portal offers two major ways to find content to watch. It provides ways to search for a movie, and it also makes recommendations based on past viewing as well as similar viewing by other viewers. According to Netflix's director of engineering, Xavier Amatriain, "Almost everything we do is a recommendation. I was at eBay last week, and they told me that 90 percent of what people buy there comes from search. We're the opposite. Recommendation is huge, and our search feature is what people do when we're not able to show them what to watch."[2] Using big data analytics, Netflix has been successfully winning its customer base from cable and telecom providers. Most Netflix customers use the cable / telecom infrastructure to connect to the Internet, and use Netflix for viewing their content.

To facilitate the development of customized content recommendation, Netflix first decided to crowdsource its recommendation algorithm during 2006–2010. Netflix made anonymized usage data available to anyone interested in competing for the best recommendation engine.[3] The competition received widespread attention from researchers worldwide, including research and development organizations, universities, and others. Unfortunately, these crowdsourced algorithms had to be stopped because of privacy concerns.[4] However, Netflix has continued to work on their recommendation engine using meta data collected from the movies and usage data collected from their viewers. Netflix employs 40 freelancers to hand-tag television shows and movies. These are product components shown in figure 4.1 above. Netflix has a team of over 800 engineers working at their Silicon Valley headquarters, developing sophisticated algorithms for combining meta data about movies with usage information from their viewers to build recommendations that viewers see on their screens.[5]

If, through interactive recommendations, a content provider can precisely measure and influence the audience, can this deep insight about the audience be shared with advertisers? Let us now turn our attention to how online advertising is changing in this interactive era.

ONLINE ADVERTISING

In the broadcast era, advertising was concentrated on a couple of media using a series of direct negotiations. Advertising agencies managed bulk purchasing of strategic spots and used their purchasing power to negotiate the best terms for their customers—the marketers. As part of their services, the advertising agencies supported marketers with creative, media planning, and research capabilities, and thereby provided a one-stop shop (see figure 4.3). Since the audiences were concentrated and the messaging was relatively unified, the ecosystem was relatively simple.

Today's viewership and associated advertising opportunities are far more complex (see figure 4.4). There are many more media formats. The display and apps change with the devices. Advertisers have linear opportunities, which are synchronized with the broadcast and nonlinear opportunities, where the viewership is for a previously recorded broadcast. Direct negotiations and bulk purchasing for advertising spots are being replaced with auction markets.

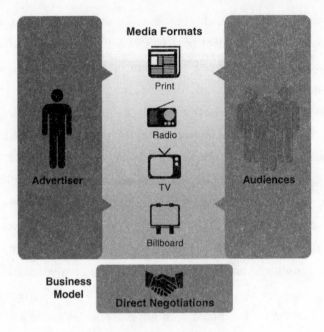

Figure 4.3 Direct Negotiations in the Broadcast Era

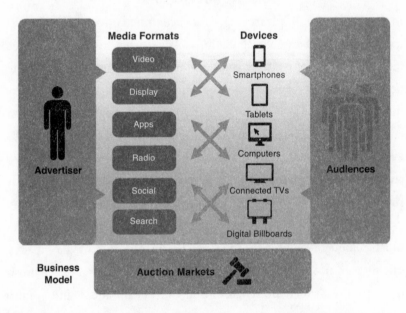

Figure 4.4 Massive Audience Fragmentation and Auction Markets

Television and radio have used advertising as their revenue source for decades. As online content distribution becomes popular, advertising has followed the content distribution with increasing volumes and acceptance in the marketplace. Digital advertising is the fastest-growing segment of the advertising business. In the first quarter of 2013, Internet advertising had already exceeded $9.6 billion, which represents a 15.6 percent increase from one year earlier, according to the Interactive Advertising Bureau (IAB) and PriceWaterhouseCoopers (PWC). "Consumers are turning to interactive media in droves to look for the latest information, to connect with their social networks, and simply to be entertained," IAB CEO Randall Rothenberg said in a statement. "This first-quarter milestone clearly illustrates that marketers recognize that digital has become the go-to medium for all sorts of activities on all sorts of screens, at home, at the office, and on-the-run."[6] Based on a study by eMarketer, per capita digital advertisement per capita spending in the United States in 2013 was projected at $201, while total media spending was $404. There are over 272 million Internet users and 152 smartphone users receiving a fair amount of attention from digital advertisers. Australia, Norway, and the United Kingdom are currently ahead of the United States in per capita digital advertisement spending.[7]

Traditional advertising was driven by reach and opportunities-to-see. A traditional advertiser could not accurately determine who saw the advertisement, and what they did when they saw it. An elaborate system of reported information was used to predict advertising effectiveness. Nielsen provides comprehensive panels today for statistically projecting television audience information and, along with their shopper survey, this information in addition to census data is used for projecting campaign effectiveness.

Google delivered a major disruption in the advertising marketplace by offering measurements and payments based on advertising clicks. Once an advertisement is placed on a browser screen, the click-rate measures its effectiveness in getting noticed by the customer. The next wave of changes came with real-time bidding for advertising. In the online advertising world, publishers such as Google Adworks offered a bidding process in which advertisers bid for placing their advertisement. In less than 100 milliseconds, Google collects a number of bids for each advertising opportunity and decides which advertisement(s) to display on the screen. Once the screen is displayed, the user action (i.e., the click) is captured and reported.

The online advertising food chain is also becoming increasingly sophisticated. The digital advertising market is rapidly moving toward real-time-bidding involving publishers, advertisers that use a complex network of demand-side platforms (DSPs), supply-side platforms (SSPs), and big data driven data-management platforms (DMPs), as shown in figure 4.5. Online advertising provides a tremendous opportunity for advertising to a micro-segment and also for context-based advertising. How do we deliver these products, and how do they differ from traditional advertising?

The advertiser's main goal is to reach the most receptive online audience in the right context, who will then engage with the displayed ad and eventually take the desired action identified by the type of campaign.[8] Big data provides us with an opportunity to collect myriads of behavioral information. This information can be collated and

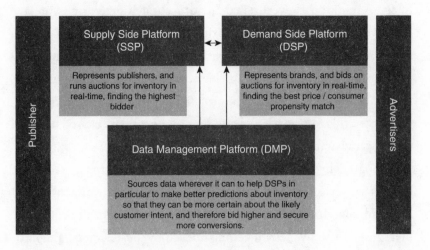

Figure 4.5   Digital Advertising Marketplace

analyzed to build two sets of insights about customers, both of which are very relevant to online advertising. First, the micro-segmentation information and associated purchase history described in chapter 3 allow us to establish buyer patterns for each micro-segment. Second, we can use the context of an online interaction to drive context-specific advertising. For example, for someone searching and shopping for a product, a number of related products can be offered in the advertisements placed on the web page.

Over the past year, I found an opportunity to study these capabilities with the help of Turn Advertising. Turn's DSP delivers over 500,000 advertisements per second using ad-bidding platforms at most major platforms, including Google, Yahoo, and Facebook. A DSP manages online advertising campaigns for a number of advertisers through real-time auctions or bidding. Unlike a direct buy market (e.g., print or television), where the price is decided in advance based on reach and opportunities to see, the real-time ad exchange accepts bids for each impression opportunity, and the impression is sold to the highest bidder in a public auction. DSPs are the platforms where all the information about users, pages, ads, and campaign constraints come together to make the best decision for advertisers.

Let us consider an example to demonstrate the flow of information and collaboration between publisher, ad exchange, DSP, and advertiser to deliver online advertisements. If a user initiates a Web search for food in a particular zip code on a search engine, the search engine will take the request, parse it, and start to deliver the search result. While the search results are being delivered, the search engine decides to place a couple of advertisements on the screen. The search engine seeks bids for those spots, which are accumulated via the ad exchange and offered to a number of DSPs competing for the opportunity to place advertisements for their advertisers. In seeking the bid, the publisher may supply some contextual information that can be matched with any additional information known to the DSP about the user. The DSP decides whether to participate in this specific bid and makes an offer to place an ad. The highest bidder is chosen, and their advertisement is delivered to the user in response to the search. Typically, this entire process may take 40–80 milliseconds.

A DMP may collect valuable statistics about the advertisement and the advertising process. The key performance indicators (KPIs) include the number of times a user clicked the advertisement, which provides a measure of success. If a user has received a single advertisement many times, it may cause saturation and reduce the probability that the user will click the advertisement.

A DMP can be effectively used to understand micro-segments and the advertising focus on these micro-segments. For example, by tracking spending on online advertising, Turn has been able to collect valuable insights about the "digital elite" segment:

> *"Marketers are beginning to understand the benefit of engaging the "digital elite" audience for their own brands and having a conversation with them across channels," says, Paul Alfieri, vice president of marketing, Turn. "In 2013, there's been a 200% increase in our customers' use of paid data to target campaigns across mobile, display, video and social, and the payoff is clear in the lift in results when they reach consumers through all the media they touch." Global marketers now have an unprecedented opportunity to reach across channels to engage in meaningful conversations with consumers moving from device to device, shifting formats and media. A recent Forrester study reveals that 90% of adults use three different device combinations to complete one simple task, such as booking a restaurant table or buying a pair of pants. And the increased time consumers spend watching videos and checking Facebook is being noticed by marketers and matched with ad spending from New York to London, from Sao Paulo to Tokyo. In 2013, we see brands ramping up quickly to keep pace with ever-moving consumers, following them across mobile, video, display, and social media.*[9]

If the DMP were noticing a new breed of multichannel shopper, how would a marketer gear up multichannel marketing to influence such shoppers? Let me use the next section to delve into the new world of multichannel shoppers.

MULTICHANNEL SHOPPING

Best Buy provided me an equally interesting shopping experience to buy a television. I shopped for it using the Best Buy website, and ended up calling their call centers to ask specific questions, which concluded the sale by phone and directed me to the nearest store to pick up the merchandise. For my family, multichannel shopping is very personalized and intent specific. My wife would not hesitate to buy electronic and household items online, but would never make apparel purchases online. I am almost always buying apparel online, because I have an odd size, and it is much easier to find my size online with a single click, instead of spending hours at a brick-and-mortar store. As we shopped for kitchen appliances for our remodeled kitchen, we often went to store showrooms and used our mobile devices to collect more information while walking through a showroom. A new study by Ipsos MediaCT and the IAB suggests shoppers are showrooming when it comes to consumer electronics, but that use of mobile phones in stores also leads to in-store sales. While 42 percent of people who used their phones while shopping ultimately made their purchase online, a full 30 percent did so in the store. Based on an online survey of 482 consumer electronics shoppers in February, Ipsos researchers found that mobile-equipped shoppers were also more likely to make an unplanned purchase: 32 percent in-store compared to 22 percent online. Nearly a third (31%) overall used mobile phones for shopping-related activity in stores. Three-quarters of consumer electronics shoppers went to a retail store to sample a product, with half planning to buy there, and a quarter intending to showroom. Digital advertising also played a role in showrooming. Over a third of shoppers (35%) recalled seeing ads for electronic products they were shopping for. Half said that digital ads made them visit an online store, while 28 percent said digital ads led them to shop at a brick-and-mortar retail location. The rest were not influenced either way.[10]

Collaboration across channels is a relatively recent development. It requires discipline in developing common offerings across channels and cross-channel training to sellers so that they can naturally work with other channels. Most traditional marketing uses channels to differentiate offerings, rather than using channel differences to collaborate for a specific selling opportunity. A recent Experian Marketing Services survey showed that just 44 percent of those asked had integrated their online and offline marketing programs. So 56 percent of the respondents' customers could be receiving uncoordinated messages or offers at the wrong time and in the wrong channel.[11] This has major implications for marketing and sales operations, as we will discuss in chapter 7.

INTELLIGENT CAMPAIGNS

Two forces come together to make an intelligent campaign. First, marketers need to define the mechanism for identifying a customer with a specific need. Second, they must present the offer at the right moment, precisely when the customer is seeking a solution that meets the need. With all the power of big data, if we can devise ways to observe the customer and deliver a campaign to the right customer at the right time with appropriate packaging, we have graduated to an intelligent campaign.

I will start with an example in which the marketer failed to deliver the campaign at the right time. I love ice cream and often go to my favorite ice cream parlor, which will remain nameless so that I can continue to receive my coupons for a discounted scoop of ice cream. The store offered to deliver ice cream coupons to my smartphone if I registered my smartphone with them. With my frequent travels, the ice cream coupon was hit-or-miss. It always carried an expiration date with a "please present the coupon by xxx date," and if I was traveling on that day, a scoop of ice cream was too far away. The worst delivery timing was when I was visiting Sao Paulo, Brazil. Not only I was too far away to use the coupon, but I also paid $0.50 in international short message service (SMS) charges. This inspired me to create a cartoon to illustrate a not-so-intelligent campaign (see figure 4.6).

By Gaurav Deshpande & Arvind Sathi

Figure 4.6 A Not-so-intelligent Campaign

The prepaid wireless market is fairly competitive in the growth markets. In most situations, consumers buy their mobile device directly from cell phone manufacturers, while telecom providers sell subscriber identity module (SIM) cards to enable these devices to use their network. If a consumer runs out of SIM card minutes, he/she looks around for a SIM card retail outlet and buys the next SIM card, which might possibly belong to another telecom provider, thereby resulting in brand switching. However, the first telecom provider has a running balance of remaining minutes for each subscriber and can possibly remind the customer to buy the next SIM card when the balance is low and there is a SIM retail center nearby. An intelligent campaign engine can keep track of location data and remaining minutes, and configure a campaign appropriately at the right moment.

I wish my car would do the same. Most of the time, I see a flag for an empty gas tank on my dashboard right after I have passed a gas station. The car has a navigation system with a full understanding of the location of gas stations. It also has an understanding of my current location and has a flag that tells me when I am running out of gas. This would require an additional capability of combining three data items from two sources, and I would be relieved of tense moments when I am praying the car will not run out of gas before I find the next gas station.

Customers may have the best intentions to use promotions targeted to them, but may not act upon those promotions. Let me take the example of coupons from the consumer products. Traditionally, these coupons were delivered in a printed form, such as Sunday newspapers, coupon booklets, and so on. Many coupons were printed, but unfortunately were not seen by customers. The precious few coupons that attracted customer attention still had a small chance of being redeemed. Customers had to cut out the coupon, bring it to the store, find a product, and redeem the coupon. There was a chance of leakage in each of these steps. Electronic coupons started to bridge the gap. Groupon offers coupons that can be organized in a mobile wallet on smartphones and can be redeemed, thereby reducing the number of items a consumer carries to the point of consumption. The use of smartphones for grocery shopping has provided grocers with the next level of automation in coupon redemption. If the store offers a Wi-Fi spot, customers' walking through the grocery lanes can be tracked, and the smartphone can be used to identify lanes where marketed products are placed. At the end of the shopping experience, the shopper can link the phone to the point of sale, offering the coupons to be automatically uploaded for redemption.

Campaign success can be traced across customers and used for fine-tuning the targeting of the campaign. It is possible to trace customers who are not likely to respond to a campaign and improve campaign yield. Also, the impact that a campaign has on customers can be studied using experiment design, as a marketer may test several treatments to different subsets of the target market, comparing their effectiveness and choosing the one with the best results.

BIG-TICKET ITEMS AND AUCTION / NEGOTIATION MARKETS

So far, most of my examples have focused on typical consumer marketing items—apparel, consumer electronics, telecom, credit cards, travel, and so forth. How is the influencing process changing for big-ticket items, such as houses, cars, and fine arts? Most of these items are specialty items that require negotiations or auctions that use one or many intermediaries. Automation and web connectivity have influenced how buyers and sellers come together and have transformed the influencing process in auction and negotiation markets for big-ticket items. In an auction market, a group of buyers and sellers settles a transaction through competitive bidding. In a negotiation market, a dealer may support the sales process. These are imperfect markets, as there could be serious supply-demand imbalances leading to significant price fluctuations, and hence opportunities for and paranoia about abnormal profits. Savvy marketers have been able to create an aura for their product and use that buzz to create extra value in the marketplace. How would big data change the influencing process for these markets, and what can marketers do to enhance their marketing using many observations?

To study the impact of new multiway communications and collaborations, and to project how auction markets would change, I decided to first look at traditional auction markets. I sat down with Ashish Bisaria to understand the auction markets for Manheim, a division of Cox Communication, which deals with the auction of cars in business-to-business markets. Most Manheim car buyers are used car dealers. They also have a regular group of sellers: mostly car fleets for large corporations and car rental companies. Manheim uses a number of car auction sites, which are gigantic parking lots, where cars are paraded at the time of auction. Traditionally, buyers purchased these cars after taking a good look under the hood. A number of visual parameters, including sight, sound, and under-the-hood inspection drove buyers to purchase specific cars. Car auctions also involved an element of rivalry among the buyers, as these savvy buyers knew each other and competed with each other for the best cars.[12]

The digital revolution is rapidly changing the traditional car auction business. Auctions are increasingly being settled in electronic markets, where the buyers offer bids electronically. Unlike traditional car auctions, these electronic auctions use digitized information about cars and seek buyers electronically. A buyer does not need to be in the same parking lot or even in the same city to participate in an electronic auction, and may not be familiar with the auction bidding behavior of the other buyers. Electronic auctions generate a larger number of buyers and more quantitative information, possibly bringing these auctions closer to a perfect market with more efficiency and lower margins for the seller. In the same dealership, a grandfather may look forward to inspecting each car under the hood before the auction and compete with others to get the best deals, while the grandson may sit behind a large screen watching several auction markets and searching for the best deals at many places focusing entirely on quantitative information provided by the suppliers.

Real estate transactions among homebuyers are a typical example of the negotiation market. Traditional real estate markets were dominated by real estate companies, such as Remax and Coldwell Banker, whose agents knew specific geographies and worked closely with the sellers to get their houses to the market. The electronic revolution, however, has brought a new set of discount brokers, such as Redfin, which work with savvy buyers and sellers to conduct these transactions. With freely available real estate information from a variety of information services providers, such as Zillow and Trulia, buyers can effectively search and short-list houses. For a buyer, much of the shopping has turned electronic, with an occasional physical tour of the house once the buyer has a serious interest.

Stock markets are the ultimate electronic auctions. Transactions are conducted using speed-trading platforms in which one second is too long. With a large number of options accompanying the stock, there is a proliferation of products, and savvy buyers use a myriad of electronic sources to collect an enormous amount of information about the marketplace before deciding on their transactions. Real-time bidding for advertisements is an equally complex high-velocity platform where trading speeds are counted in milliseconds.

The buying and selling of big-ticket items is a tricky business. The pricing is more variable because of supply-demand imbalances and the lack of a perfect market. In addition, trust becomes an important issue. Over a long period of time, a consumer may end up spending more on food, wireless devices, and apparel, but each atomic transaction is relatively small, and in general there are easier opportunities to switch. In comparison, the purchase of a house or car is a high-value transaction from an unknown source, and has significant risk potential. At a higher end, art purchases are even higher in risk. As a recent *Forbes* article relates, prices can be artificially inflated by gaming the auctions.[13] The electronic markets offer marketers an enormous opportunity to collect many observations about their customers and products, and offer them to the marketplace. These observations can be used to create a buzz or to reduce the risk associated with the transaction. They present a good opportunity for an increased flow of buyer-seller interaction information, as well as a chance for the seller to use buyer information along with product information to improve the marketing process. In the absence of the physical product, an electronic auction must include enough information either about the product or the seller's reputation, so that the buyer can make an informed judgment. In most commercial situations, such as car auctions among dealers, auctions represent repeated transactions, where the past transaction history can be used by buyers and sellers to make good decisions about the next auction.

Auction and negotiation markets are still evolving for high-ticket items. For example, in the real estate market, the success so far has been in organizing housing data for shopping, as performed by information services companies like Zillow and Trulia. Transformation efforts, such as Redfin, are still struggling for acceptance in the marketplace.[14] Marketing will evolve with the auction markets and will start to utilize the rich collaboration possibilities for improved product marketing.

GAMES, VIDEOS, SMARTPHONES, AND TABLETS

When personal computers were introduced in the 1980s, the most common consumer application was word processing. Millions of consumers abandoned their typewriters to move en masse to sophisticated word processors, which provided them far more productivity. When tablets were introduced, a similar killer application emerged to make them popular. Unlike word processors, which improved productivity, "Angry Bird" and other games were productivity killers among executives. I watched in dismay as my colleagues purchased expensive iPads so that they could conquer the angry birds and invested countless hours in perfecting their skills. Unlike the "punch card" generation to which I belong, these young executives had personal computers when they went to school and had grown up playing hangman and solitaire on them. Tablets gave them the screen size, the power of graphic animation, and a set of games that attracted and held their interest.

Games are far more sophisticated today than the earlier Nintendo games my kids grew up with. My nephew proudly demonstrated the power of collaboration as he downloaded his favorite games on my tablet and his own. Within minutes, we were busy on our respective tablets shooting bullets at each other, using my Wi-Fi to connect the two of us. He showed me numerous ways in which I could collaborate with many others, friends or strangers. If I could let a stranger shoot bullets to my second life going through its third incarnation, why could I not use the power of graphics to sell product features and demonstrate use cases?

Tablets are offering new avenues for a variety of graphic applications. Video content is rapidly finding its tablet audience. I often watch *Dancing with the Stars* on the ABC app on my iPad, freeing me from the proximity to my television or the prime spot in my day when I am quietly resting on a flight at 35,000 feet or busy with other activities. The show requires me to repeatedly view commercials at regular intervals, which I am not allowed to fast-forward through. I have also found that different apps have different ways of dealing with my preferences. Some ask for a "like" vote to figure out what type of commercials would be of interest to me. Some offer to skip the commercial after a couple of seconds, but if I decide not to skip, the commercials last a lot longer than the customary 20 or 30 seconds.

Videos and games on tablets provide us a new set of capabilities for influencing customers. We have barely scratched the surface. Unlike a television, the tablet is often associated with an individual and is frequently connected to the Internet. We are more likely to carry a tablet with us to malls and stores, especially while making purchases of expensive items. We also use tablets for video calls and video games.

With a large number of observations, and having the knowledge of tablet location, marketers can use tablets for powerful context and intent-specific messaging. This year, as I prepared to watch my favorite tennis tournament—the US Open—I downloaded the US Open app on my smartphone. The app provided me with a lot of information at silicon speed, including player statistics.

CROWDSOURCING AND JAMS

The Internet has provided an avenue for billions of individuals to interact with each other. If I ask for an idea, I can receive millions of ideas, and if there is a way to organize crowds, a marketer can use his/her customer base to provide them with valuable insight that would be hard to collect otherwise. Sometime work is performed with a small incentive, and often with minimal payments by the organization receiving the service. For example, if I am looking for an illustration or a photograph, I can pay a professional photographer $100 to $150, or else get it from iStockphoto for $1. iStock provides royalty-free stock photography, clip art, vector illustrations, and audio and video clips that are used by businesses and individuals around the world in a wide variety of projects.[15] iStockphoto collects these artifacts from a community of contributors around the world. Using Powerpoint and iStockphoto, I am able to create professional-quality presentations.

Jeff Howe first used the word "crowdsourcing" in an article for *Wired* magazine.[16] He restated the definition in his blog on Typepad:

> *Simply defined, crowdsourcing represents the act of a company or institution taking a function once performed by employees and outsourcing it to an undefined (and generally large) network of people in the form of an open call. This can take the form of peer-production (when the job is performed collaboratively), but is also often undertaken by sole individuals. The crucial prerequisite is the use of the open call format and the large network of potential laborers.*[17]

Marketers are using crowdsourcing for many tasks, which would have been impossible earlier. One of those tasks is to collect product ideas. As I was composing this chapter, a fellow traveler sitting next to me on the airplane started a very interesting discussion with me. He told me about a website for the US Army, armycocreate.com, which facilitates our armed forces' use of crowdsourcing as a way to get ideas to the army. With proper governance, crowdsourcing can be a good source for product development and marketing. The website armycocreate.com provides a number of ideas and outlines a process for ideation, solution development, and testing.[18]

A related mechanism for collecting innovation ideas is Collaborative Innovation™.[19] In a world where innovation is global, multidisciplinary, and open, it is necessary to bring different minds and different perspectives together to discover new solutions to long-standing problems. Therein lies the essence of collaborative innovation. IBM's jams and other Web 2.0 collaborative mediums are opening up tremendous possibilities for collaborative innovation—ways of working across industries, disciplines, and national borders.[20] The most noteworthy jam, organized by the government of Canada, the United Nations, and IBM, is the "Habitat Jam," which was intended to conduct an Internet dialogue on sustainability and which attracted 39,000 contributors from 158 countries.[21] IBM conducts a number of jams with employees and customers to collect innovation ideas, and has used these jams to gather ideas for new business ventures.

With the help of all the big data sources described in chapter 3, marketers can get a good handle on product usage. The Achilles heel for marketing is our ability to collect observations about what does not work. Researchers have used extensive surveys and consumer panels to gain a better understanding of new product ideas from customers. Crowdsourcing is comparatively much cheaper and often results in good ideas.

ENDORSEMENTS AND VIRAL BUZZ

Google India posted an advertisement on YouTube on November 13, 2013. Here is the gist of the advertisement: a man in Delhi tells his granddaughter about his childhood friend, Yusuf. He has not seen

Yusuf since the Partition of India in 1947, when India and Pakistan became separate countries, and the two friends were forced to separate. The man's granddaughter arranges for the two to meet again.[22] It is a 3 minute, 32 second advertisement that would be considered too long for a conventional advertisement. It shows the Google products being used in a "use case," and it attracted more than 3 million viewers in the first three days it was posted.

Google posted the video as a content to share. A number of blogs and news items covered the story in the next day or so, and the stories started to appear in other social media sites. As viewers watched the video, they felt touched by it and started to share it in their own personal pages, each of these links reinforcing the traffic to YouTube, where the original video was posted. This is a great example of endorsement and viral buzz. It would take an enormous advertising budget to hit 3 million views of an advertisement in the first three days in traditional media. It is also important to note that advertisement content is driving the viral buzz. Google India has produced several advertisements in the past, mostly communicating product information.[23] While the story buildup in the reunion advertisement took a lot longer, it is immensely more popular than the straitjacket advertisements.

Consumers have already discovered social media as their platform for sharing product information. In that case, how would a consumer deal with a poor service quality experience? An IBM Global Telecom Consumer survey was conducted with a sample size of 10,177.[24] In this survey, 78 percent of the consumers surveyed in the mature markets said they avoided providers with whom friends or family had had a bad experience. The percentage was even higher (87%) in growth markets. In response to a related question, they said that they informed friends and family about their poor experiences (73% in mature markets and 85% in growth markets). These numbers together show a strong influence of social networks on purchase behavior. These are highly significant percentages, and are now increasingly augmented by social media sites (e.g., the "Like" button placed on Facebook). The same survey also

found that the three most preferred sources for recommendation information are the Internet, recommendations from family / friends, and
social media.

In response, marketers have started to embrace expert advisers to
support their claims. Amazon has one of the most elaborate reviewer
networks. Each reviewer is rated based on the number of reviews he/
she has done and how often they were read, liked, and used by others
in their purchase decisions. It is an honor to be ranked high on their
review list, and Amazon offers that information as a way to empha-
size the credibility of the reviews. Lisa Mancuso, senior vice president
of marketing for Fisher-Price, talked about the company's ambassador
program in an interview with *Forbes* magazine: "We know that more
than two-thirds of mothers consider blogs to be a reliable resource for
parenting information, so we have created a robust program to connect with parenting bloggers around the world. We call them our Play
Ambassadors."[25] Such programs, when actively integrated with social
media accounts, give organizations the capability to start differentiating
themselves in their ability to converse with customers.

The buzz created via blogs, both positive as well as negative, can
be measured and used for fine-tuning a product or messaging. I will
discuss in chapter 7 how marketers are using social media command
centers to collect, analyze, and act upon the social media activities associated with their brands.

PROPOSITION

In the previous chapter, I described several sources of big data that
could be used for gaining a better understanding of customer. In this
chapter, I have provided several examples of how marketers can influence their customers. I remember talking to my aunt a long time ago,
who was disgusted with advertisers. I politely told her I worked for
one, and she responded, "It is a bunch of lies." She was reacting to
a set of broadcasted messages coming from marketers to consumers. The messages were shamelessly repeated until the consumer

finally remembered them, and marketers hoped the consumers would remember them at the point of purchase. The credit card offers described at the beginning of this chapter are another good example. As long as the yield is above zero, marketers have a justification to do push advertising.

Contrast this with the new way of marketing by Chipotle. Any time they open a new store, they distribute numerous coupons to the residents in the neighborhood, offering them a free meal. Their assumption is that the meal will be so good that customers will not only repeat their visits but will also tell others, and this form of word-of-mouth publicity only costs them a lot of free burritos on the first day of opening a new store.

As consumers become better organized, they may seek the opinion of others to decide on their brand purchases. While consumer needs are individual, we live and work in a collaborative society where we feel connected with people with similar needs, wants, or likes. I am amazed at the collection of books Amazon is able to offer to me based on the books I read. This is a very **interactive** process. The more I use the network to buy additional books, the more the recommendation system develops stronger ways to gather a group of customers with similar reading preferences.

Jeff Jonas (an IBM Fellow and a leading expert on big data) once told me the real power of big data is that it finds new data. Questioning the source and getting more focused data is the best way to fill in the gaps. But doing so means that we must have earned the customers' respect. Once a channel of communication is established, the marketer can use the channel to encourage customers not only to buy products, but also to collaborate in bringing other customers to the marketer. Once marketers have found a consumer with a need, they can use other observations and interactions with consumers to get a better understanding of the need and start creating a community around the buyer that the he/she will trust.

Our decisions are colored by the information available to us and by the opinion of those who matter most to us. Hopefully, this chapter is consistent with your personal observations and has convinced you that marketers have available to them a set of sophisticated capabilities for influencing their customers in the most personalized and collaborative ways. Marketers can use these capabilities for marketing research, pricing, and the promotion of their products and create a truly individualized experience for each customer based on his/her preferences and needs.

5

FROM SILO'ED TO ORCHESTRATED MARKETING

INTRODUCTION

So far, I have explored the power of observation and the ability to collaborate with customers. But how do we convert all this into a razor-sharp focus on a specific set of customers? The marketer has now the opportunity to use this power to bring the customer to a positive decision about a product, whether this is a first-time purchase, a repeat purchase, or a tweet to friends exalting the virtues of the recently purchased product. These decisions happen over time and require a series of collaborations. Without a proper conductor, the musicians hired to influence the customer can at best create musical noise. How do we orchestrate these powerful tools to collaborate with each other? This chapter discusses how marketing efforts can be pooled across the silos to influence a customer through the stages of marketing. How would marketers coordinate the effort to reduce cost and the annoyance factor and use the power of collaboration to improve the relationship with their customers?

I poured through online advertising information and found that many large telcos invest tens of millions of dollars in advertising with Google. For example, AT&T invested $40.8 million in Google advertising, and Verizon invested $22.9 Million in 2011.[1] According to the Interactive Advertising Bureau (IAB), telcos invested 12 percent of the

$20.7 billion in overall US online advertising in the first half of 2013.[2] The payments are made either directly to Google or through the real-time bidding system operated by advertising agencies, demand-side platforms (DSPs), and supply-side platforms (SSPs) to purchase online spots from Google. Presumably, much of the advertising happens through the communication infrastructure provided by these telcos. I was curious about how much of this advertising was directed to an existing customer and using a basic advertisement. Telco marketing to a customer should know for certain who the customers are, and be able to refine the advertising based on what the customer has already purchased. Given that Google offers advertising opportunities to a DSP, is there a way the advertising agencies can direct the DSPs to bid for different advertisement based on customer status? Can marketers focus their advertising based on where the customer is in the buying cycle? As I talked to a number of telecom providers, I found they are in various levels of maturity in targeting their advertising to their customers, based on past purchases and current customer interest as observed through their browsing behavior.

The orchestration requires a couple of components, which can be shared across these organizations. It also requires a clear navigation through an external exchange of information, keeping in full view privacy policies and differences in customers' privacy preferences. Some of the challenges in orchestration are organizational. Marketers need to fully understand the data bazaar introduced in chapter 2 (and explained in more detail in chapter 7) and how each player is economically motivated to participate. This new infrastructure tears apart the advertiser, agency, publisher, and media research network of the past and puts in place a new ecosystem led by information services giants like Google and Facebook. This chapter will describe technological, organizational, and legal / regulatory issues faced by marketers and how orchestration is being achieved by the pioneers.

Let me focus on the "work-at-home" customer whom I discussed in the previous chapter. Through analytics, a marketer finds a segment of telecom customers who are in need of big bandwidth during the day-time, in a residential location as they work from home. How do we identify members of this segment? How do we track each member of the segment as we organize marketing campaigns? How do we keep track of someone who is responding favorably to the campaign and could easily be an early adopter of the program? How do we accelerate the campaign to support shopping and selling to the early adopters? Can we seek the help of early adopters to attract their social group and influence others to follow suit?

CUSTOMER PROFILE

As I answer these questions, the first obvious orchestration asset is the customer profile. In the broadcast days, marketers treated segments of customers, established marketing programs to reach each segment, and provided channels to support these segments without an explicit need to record the status of the campaign with each customer. Now, as we turn from broadcast to collaboration, the first question is, whom to collaborate with? A customer profile provides us with a list of prospects. As we proceed with collaboration, we must keep track of each of these customers and track their buying process by keeping track of their needs, their response to marketing campaigns, and their current understanding of our offerings.

In any enterprise, there are likely to be many views of prospects and customers. Most of the fragmentation comes from divergent views across organizations. While customer care organizations focus more on customer interactions, billing organizations track billing information. If my son and I are sharing a customer account, customer care may have knowledge of places he visits, as in service addresses, while a billing organization may only care about his billing address, where they send the bill. Customer master data management (MDM) solutions are popular ways of bridging views and bringing together a single unified view. However, over the past decades, this integration has been focused primarily on intraorganization sources of traditional "structured" data.

Most often in a structured data environment, the data must be integrated across a couple of organized sources, each carrying a customer ID as well as a customer hierarchy. As different sources are combined, they represent their IDs at different levels in the hierarchy and carry other information about the customers that can be used for merging customer data. For example, a customer care system that collects web clickstream data may deal with an individual customer and carry his/her name, user identification, location of interaction, and so forth. The data in the billing system, however, may carry an account identification associated with a billing account for a household and the household billing address. A unified customer profile may carry many customer IDs from a customer care system mapped to a single billing ID from the billing system and, depending on the analytics requirement, the appropriate data can be organized and summarized from either of the two sources.

Usage data, which is described in chapter 3, provides the dynamic extension to the classic MDM-type solution. Unlike structured data, usage data may be used to create a number of customer attributes, which may change over time. For a quick-service restaurant, such as Starbucks or Panera Bread, loyalty card data can be used to identify a large number of customer profile attributes. These attributes may include usage preferences, locations, response to promotions, local weather patterns, and so on. In the location analytics example, I discussed mobility patterns, such as "work at home." While the raw location data associated with a subscriber is fairly structured, these attributes and related microsegments are relatively more dynamic. Most marketers offering loyalty cards have begun to mine usage data to align additional customer attributes to static customer information available from customer service and billing. By adding usage data, we can start to differentiate these customers based on product usage, service location, time of day, day of week, or other significant attributes of interest to a marketer.

With the wide availability of social data, we have opened up the customer profile to also take into account social sources, including

Twitter, Facebook, Yelp, YouTube, other blogs, and in general any information that is publicly available. The information published externally could include intent to buy, product preferences, complaints, endorsements, usage, and other useful segmentation data. This data can be collected, collated, and identified with individual customers or segments, and connected with the rest of the customer view. How do we merge internal and external views to create what we may call a big data view of the customer? This integrated view is a far more holistic understanding of the customer. By analyzing and integrating this data with the rest of the customer master, we can now do a far more extensive household analysis. This data may reveal additional information about customer satisfaction with the product. For example, there may be low usage for a product because of lack of access, as opposed to disinterest in the product. The social media chatter may be able to discern geographical locations where scarcity is leading to lack of product usage. While this data can also be collected via better data collection at the point of sale or through consumer surveys, social media data collection may provide a more comprehensive sentiment at a lower cost, and could be far more dynamic in revealing spikes in sentiments and associated causes.

Another good source of data is customer profiles from other industries. As a retailer, I may have a good customer profile about my customers' usage pattern. A telecom provider may provide complementary understanding of their mobility patterns, and a cable operator may have a good understanding of their media viewing habits. Mobility patterns are key to providing context-specific promotions to customers. For example, while the parents may be paying for the cell phone, the actual user may reside in a college dormitory in a different city and should not be offered promotions for regional stores in the city where the parents live unless the student is visiting the parents for Fall break. Figure 5.1 shows sample elements of a big data customer profile. It includes demographics, social patterns, buying patterns, and mobility patterns. How would a marketer find these jigsaw pieces and pull them together to get a holistic view of the customer?

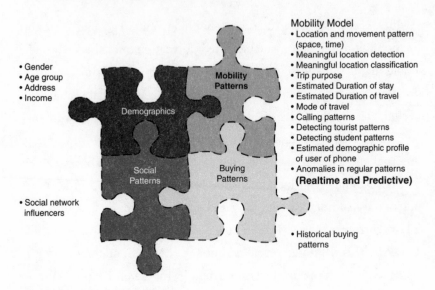

Figure 5.1   Big Data Customer Profile

A number of cloud-based marketing organizations are keeping track of web usage patterns. By analyzing the websites browsed by a specific customer, these organizations are establishing their customer profiles and creating attributes like "interested in golf," "stock investor," and "classical music lover." Data management platforms combine this data with the past history of advertisements placed, viewed, and clicked, to generate a sophisticated understanding of a customer and his/her interest in a specific campaign, and the saturation level.

The customer profile must stay focused on its purpose. For example, the Obama campaign created a voter profile with two objectives—one to predict the likelihood of someone voting for Obama, and the other to predict the likelihood of someone contributing to the Obama campaign. This campaign began in 2012, the second-term election year, and tracked the name of every one of the 69,456,897 Americans whose votes had put him in the White House in 2008.[3] They may have cast those votes by secret ballot, but the analysts could look at the Democrats' vote totals in each precinct and identify the people most likely to have backed him. They started with a customer profile for

180 million potential voters and updated their information on a weekly basis to drive a series of electoral campaigns.

ENTITY ANALYTICS

We now have an interesting challenge. There are several sources of customer data. This data may be generated at different levels in the hierarchy. How is this data aligned across sources to create a unified customer profile? For example, the cable provider may provide channel-surfing information for my household, while the wireless provider may provide mobility patterns for my son, who does not live with me, and hence does not participate in cable viewing in my living room except when he visits my house. However, he shares a family contract with me, and shares the billing address. Each of us tweet to our respective social groups using a Twitter handle and use endomondo to post our bike riding and jogging records in Facebook. His tweet and jogging locations do not correlate with his billing address. At the same time, our family vacation brings us together, and a marketing campaign for travel spots can be directed to either of us. The telecom provider may be willing to sell the mobility data to the airline interested in offering me travel deals. However, the telecom provider may only provide the data at an aggregate level for 25 or more customers in a micro-segment to protect individual identities. How should a marketer organize and align this data? In order to have a meaningful dialogue, a marketer must bring this data to a unit, which can be related to a marketing action. For a television advertisement on regular television, that unit is the household. To a multidevice consumer, the unit could be a specific device used by an individual. The marketer needs a common denominator and an aggregation mechanism to roll up or down the hierarchy. In the case of the Obama campaign, the individual voter data was periodically aggregated for television media-planning decisions.

Customer profiles have been a subject of focus for decades. Customer relationship management (CRM) tools were the first to offer an integrated customer database, one that would unite sales, revenue, and services views of a customer. Aaron Zornes has been a father figure to the customer data integration / master data management (CDI /MDM) groups and has provided a much needed backbone to this area with his MDM Institute.[4] His graceful beard is gradually turning white as he patiently tracks the progress of the MDM community. They did a fair amount of pioneering work in tearing down the organizational silos. As organizations merged and departmental information technology (IT) investments were centralized, they found that each organization had a different view of the customer. For instance, in a newly merged diversi-fied insurance organization, the health insurance department tracked various health stages of a customer, while the life insurance department cared about only two issues—whether the customer was alive or dead. The telco provisioning department carried 96 states of customer order, while the sales team had only 6. The billing team was focused on the billing address, while the trucks were rolled out to the service address. To make the situation more confusing, all of these organizations used the same words—customer, product, address, order, and so forth to mean different terms. The first attempt through corporate initiatives, driven by regulators, such as the Sarbane-Oxley Act on corporate reporting compliance,[5] was to establish a centralized data model that served everyone. While the attempts succeeded partially, they resulted in models that were hard to change and too static for most businesses. Gentler approaches using registry-style ID mapping or coexistence MDM, in which master data was consolidated as needed, found more popularity than consolidation MDMs, in which a central customer master supported diverse needs.[6]

The central technical capability in any MDM is its ability to match identities across diverse data sources. How do we integrate big data with the matching capabilities of the MDM solution? Most MDM solu-tions offer matching capabilities for structured data. MDM software matches customers and creates new IDs that combine customer data from a variety of sources. These solutions are also providing significant capabilities for using customer hierarchies to normalize data across systems. However, in most cases, the format for the data is known, and the content is primarily structured. What does source data for big data's single view of a customer look like?

Blogs and tweets posted by consumers on social media sites provide a wealth of information for sentiment analysis; however, this data is not structured. Consumers do not always use proper company or product names. The data contains a fair number of slang words, and there is a mix of languages in a multiethnic, multilanguage environment. Consumers may use a variety of words to convey positive or negative sentiments. The link to the author is not very well articulated. Therefore, analysts start with scant information, such as Twitter handles and unstructured references, and filter and link this data to decipher demographics, location, and other important characteristics required to make this data meaningful to a marketer. At the end of the day, the social media data hangs from its own customer ID, which can be aggregated or abstracted by a marketer.

In chapter 3, I covered several other sources of observations that provide a wealth of customer data. However, customer data belongs to one industry and must be filtered or transformed before it is used by another industry. Let us consider the example of location data. For a wireless company, location data may include its source—whether cell tower, Wi-Fi, or Global Positioning System (GPS) data collected from a device. This would be very useful for network performance analysis, but has limited value outside the wireless industry. Possibly, a marketer would be interested in a polygon that defines a geographic boundary, time duration, and a list of people who were inside the polygon in the specified time period. If the polygon represented a mall, the marketer could assign a context and use the data for a variety of purposes.

The hardest job in data sharing is to come to an agreement on the definitions and to align collaborating parties to the same definition of the data being exchanged. A marketer in a telco may find ways to mine location data and provide a large number of new attributes, which would be of tremendous interest to a consuming industry. However, now we are talking MDM data-governance issues on a much larger scale than ever before discussed or tackled by the MDM industry. What if a telco can track web-browsing activities by location and turn that data into micro-segments, with the ability to identify hockey fans visiting a specific sports bar, and package this data for a marketer who is selling sporting goods near that location? How do we govern this data so that the two industries can relate to each other's definitions without creating a static model that cannot be changed with the new fashion trend?

This data coming from a variety of sources could be linked together to get a holistic understanding of the customer. Privacy laws are still evolving in understanding what data can be linked, with or without customer permission. Identity resolution is the next step in the evolution of matching technologies for MDM. Jeff Jonas has been working on IBM's entity analytics technologies. His initial work was for the casino industry, where he developed technologies to identify casino visitors who profited through fraudulent gambling. This is a powerful technology that takes into account both normal as well as deceptive data from customers.[7] The technology is based on a set of rules that places the probability of a match on a set of seemingly unrelated facts. As hard facts match, the probabilities are altered to reflect newly discovered information. Customer-initiated actions, such as accepting a promotion, can be hard evidence added to customer handles or user IDs, connecting them to device IDs, product IDs, or customer account information. Jonas has been studying identity resolution in a number of big data entity analytics problems, including the US voter registration process.[8]

Customer data can be organized at different levels of hierarchy. As organizations begin to share this data with outsiders, they may restrict data access to a certain level of customer hierarchy and may share selected attributes at each level. For example, a telco may share their mobility patterns for a community, specifying the percentage of a community that travels by bus to work, without specifying the members of the community and their specific mobility patterns. While a community mobility pattern is very useful information for marketers, individual data could be both intrusive and ineffective. Let me discuss the example of using location-based mobility pattern data for targeted advertising. A telco would share community patterns, which can be used by a DSP to decide which advertisement to place to a community, for example, putting greater emphasis on leisure travel promotions to a "globe trotter" community. If the consumer gives approval, it may be appropriate to target daily promotions for nearby restaurants that are in close proximity to the consumer's most frequent hangout.

Whenever I have made this idea part of a presentation, I have seen several raised eyebrows and been asked questions about customer privacy. Customer privacy is always an area of major concern. For years, corporations collected all types of privacy information and matched it from a variety of sources to obtain a single view of the customer. However, most of that information collection was transparent to the customer and happened without full disclosure. Now, however, big data has the potential to correlate data across industries and across sources far more extensively than in the past. As a result, privacy is a major issue that I address in the next section.

PERSONAL PRIVACY PREFERENCE MANAGEMENT

The first part of the solution is a data obfuscation process. Most of the time, marketers are interested in customer characteristics that can be provided without privately identifiable information (PII)—that is, uniquely identifiable information about an individual that can be used to identify, locate, and contact that individual. All PII information can be destroyed, while still providing useful information to a marketer about a group of individuals. Now, under "opt-in," the PII can be released selectively on a need-to-know basis.

As I worked on the data obfuscation process, I found that this process is significantly more complex than expected. While PII data is destroyed, I cannot leave related information that, if joined with obfuscated data, might lead back to the individual. For example, if I destroyed the address and phone number but left location information, someone could use the location information to establish the consumer's residential address. Also, there are grades of PII information. Zip+4 or county designation may be an appropriate locater unless we are dealing with the home addresses of billionaires. Also, small samples are a problem. In the case of the Netflix matching engine, someone was able to determine the identity of an individual based on that person's media content viewing preferences and patterns in anonymized data.[9] The non-PII information can uniquely identify an individual if only one individual meets the profile. IBM has been investing in data-masking products and processes, that allow us to systematically identify PII information in a data set, tag it, select masking algorithms, test the masking process, and establish the effectiveness of the masking solution.[10, 11]

While one can reasonably expect data masking to obfuscate customer identity, it should be usable for analytics. The algorithm should remove or randomize PII, but not destroy the statistical patterns required by a data scientist. For example, if I take a set of real addresses and replace them with XXX, anyone looking for statistical patterns along geographical boundaries would not be able to use the obfuscated data. Patented algorithms examine data masking for a large database, and are able to successfully mask the data across a group of data items. The algorithms systematically work on a group of fields to destroy privacy, while leaving the data characteristics for its intended task.

A privacy infrastructure provides the capability to store information about "opt-in" and use it for granting access. Anyone with proper access can obtain the PII information, as granted by the user, while others see only obfuscated data. This solution provides us with enormous capability to use statistical data for a group of individuals, while selectively offering "one-to-one" marketing wherever the consumer is willing to accept the offers.

An audit can test whether the obfuscation process, algorithms, and privacy access are working properly in a multipartner environment in which third parties may also have access to this data. If properly managed, the data privacy framework provides gated access to marketers based on permission granted by the consumer, and can significantly boost consumer confidence and the ability to finance data monetization.

Data privacy concerns are changing with time and the generations, leading to significant differences in personal privacy preferences. A trustworthy marketing program will build its trust with a customer gradually and with a full understanding of customer preferences.

DYNAMIC PRICING

Pricing has been a hotly pursued topic for marketing, as every percent increase in price without a corresponding demand decrease means an increase in profits. However, over the past decade, we have seen a new, dynamic pricing equilibrium favoring customers and fueled by third-party pricing search tools. For example, in the travel industry, travel sites such as Bing, Travelocity, or Priceline offer customers the ability to search across an entire market and find deals. Wherever the product is perishable—theater tickets or airline seats, for example—the new marketplace lets buyers find deals for items that need to move off the inventory or else will remain unsold.

Let me take the example of Bing and discuss their ability to monitor travel pricing, and advise customers regarding timing for ticket purchase. They purchased a company called Farecast in order to include a fare history analysis as part of their travel website. For travel between two cities, Bing provides useful information on price fluctuations in the past, and based on this historical data, it predicts whether a consumer should buy a ticket now or later. I ran the advice for a trip from Los Angeles to Denver for the first week of November and received the advice to wait, as well as a prediction that the price would further drop by $50. Airline pricing data is both dynamic as well as public. Competitors and buyers can collect, organize, and analyze this data for their price optimization.

EBay and other auction marketplaces have brought buyers and sellers together where price is determined through a bidding process. This creates an interesting dynamic in that suppliers are using sophisticated price optimization models to determine price based on supply and demand, and consumers are using third parties to find the best deals. Prices change often, with constant fine-tuning based on supply and demand. These models are also bringing a new set of retailers who provide a layer of customer interface between the customer and the supplier. The retailer and the supplier now share responsibility for the customer experience, and they each have a direct impact on the resulting customer experience as felt at the end by the customer. For instance, after years of dealing with my favorite rental car company and receiving excellent differentiated service, I used a retailer and ended up purchasing a package that included travel, hotel, and car rental, only to find that the premium customer policy for car rental cancellation no longer applied to this bundled package.

ORCHESTRATION FOR CONTEXT-BASED ADVERTISING AND PROMOTION

As online advertising becomes integrated with online purchasing, the value of placing an advertisement in the right context may rise. If the placement of an ad results in the immediate purchase of the product, the advertiser is very likely to offer a higher price to the publisher. DSP and DMP success depends directly on their ability to track and match consumers based on the consumers' perceived information need and their ability to find advertising opportunities related closely to an online sale of associated goods or services. If the consumer has already purchased a product and is no longer shopping, it is a poor investment in advertising.

Let me cite a specific example to illustrate the point. Financial services and telcos are heavy spenders in online advertising. As I studied their spending, I found that in investing their advertising dollars, they are seeking specific audiences. Advertising agencies employ DSPs to participate in the bidding process for an exchange or Ad Network to place the advertisement. However, more than half of these advertisements are placed on the browser screens of customers who have already purchased the advertised product and are unlikely buyers. In many cases, the customer may be using the product, for example, the broadband connection, from a telco to view the advertisement (for a broadband connection from their current supplier). If this situation reminds you of all the mail you received from your telco offering you a bundled product of wireless, wireline, and cable, while you were already using a bundled product, you are not alone. In addition, the advertised product may not be available in your neighborhood.

So, how do we insert the knowledge of customers' existing products, their communication needs, and the product availability in their area? A telco can build a telco profile that represents their customers and their needs. They can supply this information to a DSP like Turn, which uses this information, along with online bidding rules, to identify customers with specific needs and bid up or down a telco advertisement based on the telco profile. In many situations, a marketer may have additional campaigns, such as upsell of options available to those customers.

If marketers can establish a personalized advertising connection with customers, how about taking the next step in personalization? A marketer can establish a dialogue with the customer and use advertising to make information available as it takes the buyer through the stages of buying. Online advertising can be highly personalized, as a marketer can target a personalized campaign to a specific subscriber. Let me use a scenario to show how advertising in online advertising can be personalized to the context and state of selling.

Linda is currently shopping for a smartphone. She uses a search engine to look for Android phones and finds the website for a wireless service provider that is offering Android phones. Linda is price conscious and lives in an urban area with spotty wireless coverage. While she uses the website to browse for phones, she does not pay attention to coverage and price plans. As the wireless site reports Linda's use of the web page, she is flagged as a potential phone buyer. At the same time, the next advertisement from the wireless service provider emphasizes wireless coverage and price plans—two important aspects that will move the wireless service provider closer to her buying consideration. As the wireless service provider provides an indication to their DSP, they may provide a higher bid for coverage and price plan campaigns, refining the messaging to Linda and making it relevant. The campaign may differ across subscribers, reflecting their buying criteria. In addition, the DSP would use additional criteria to decide how to bid for an advertisement targeted to Linda. If Linda were offered the price plan advertisement five times and she did not click it, the DSP might stop bidding that specific advertisement.[12]

How do we get this orchestration to work? A marketer would provide a list of target customers and the advertising campaigns to its DSP. The DSP would prioritize its bidding for advertising using the targeted customer list from the marketer, combining it with the DMP information collected regarding past advertisements to a specific customer. A sophisticated set of bidding rules would watch over the bidding process to avoid saturation and other real-time-bidding considerations. Once the customer sees the advertisement and clicks on it to purchase the product, the marketer would update the target list to exclude the customer, thus removing the candidacy of a customer from receiving future advertisements for a product he/she has already purchased.

For this orchestration to work, the customer IDs have to match. This is easier said than done. The marketer must organize its customer profile to identify an individual based on his/her browser IP address. Once the proper ID has been associated with a customer, the target list can be consumed and used for bidding by the DSP.

If we now move from advertising to promotion, through an active discount coupon, privacy rules may need to be enforced. Most marketers would only place a coupon once the consumer has agreed to receive context-based coupons. Let me take you through an example of how a marketer would engage a customer in a targeted coupon program and how the targeting would be done.

Cuppa Heaven is a new barista chain offering coffee in its retail outlets. Cuppa Heaven is interested in operating coffee stores next to movie theaters in the mall. In its research, it has found that moviegoers are likely to bring a latte to watch a movie. Cuppa Heaven would like to cater to this growing segment of the market. To identify the target population, Cuppa Heaven connects with OfferTel, seeking mobility patterns for OfferTel subscribers. Using the raw location information from its network data, OfferTel provides a report to Cuppa Heaven highlighting the percentage of subscribers from each community surrounding a mall, who regularly go to that mall at least once a week to watch movies. Cuppa Heaven decides to target the top 25 percent of the communities, sending a letter in the mail to each resident in those communities asking if they would be interested in downloading a free movie guide app. The app provides the resident with a schedule of movies at the movie theater and also lets him/her purchase movie tickets, watch movie trailers, and get promotions. Cuppa Heaven also places a poster at the movie theater offering the app to download. Based on the download statistics, Cuppa Heaven decides that placing the poster at the movie theater is the best way to promote its app.

When the app is downloaded, it seeks the customer's permission to connect to his/her Facebook and Twitter accounts in exchange for a Cuppa Heaven promotional coffee. It also offers a coupon for a bundled coffee and a movie ticket package for two. By analyzing the tweets, Cuppa Heaven decides to offer an endorsement program to have the customers place "like" on its Facebook page. All of these campaigns are analyzed for their effectiveness and fine-tuned.[13]

The above scenario is effectively utilizing an orchestration framework to analyze data, target customers, offer campaigns, compare responses, and make adjustments. The initial data from OfferTel is big data covering the entire population. The location analytics is conducted at an aggregate level, so the analysis results can be sold to Cuppa Heaven without any loss of consumer privacy. The offers to customers are based on an opt-in process, which explicitly seeks permission before making location-specific offers. The campaign's yield can be analyzed. For example, Cuppa Heaven may send different bundles to different customer sets and compare the results to seek the best bundle. This is a very closed-loop process.

CROSS-CHANNEL COORDINATION

So far, I have described advertising and promotion activities separately. However, the orchestration engines can work across channels. The recent trend is for customers to use multiple media simultaneously to connect with marketers. In return, marketers are also using multiple channels to connect with their customers.

For its eighteenth birthday, P.F. Chang, the gourmet Chinese restaurant chain, decided to use a campaign that consisted of giving away its signature lettuce wraps. The P.F. Chang promotion was conducted using Facebook, and it offered coupons for lettuce wraps. The strategy helped double its fan base, and the company saw a more than 3,000 percent increase in engagement on its page.[14]

The media spend across channels has always been closely managed by marketers, but how do we now organize and measure the collaborative forces across channels? There are three levels of coordination across channels.

The first level of coordination is associated with an awareness of customer cross-channel activities. A number of analytics programs have begun to store, correlate, and analyze activities across channels. For example, Clickfox analyzes customer events across channels to develop a holistic view of the customer experience.[15] While most of these activities have been focused primarily on care as opposed to marketing and sales, most of the techniques, instruments, and findings are equally applicable to marketing. In a typical cross-channel customer experience analytics, all customer events are collated and stitched together to formulate a comprehensive customer status. Each channel contributes its partial view of the customer. By combining this data across channels, the marketer is able to obtain a comprehensive view across all the channels. This revised understanding of the customer can now be shared with each channel.

The second level uses a channel to complete the task for another channel. The P.F. Chang Facebook lettuce wrap coupon is an example of using more than one channel to carry out a campaign. The purpose of the campaign was to make its customers aware of the restaurant's eighteenth birthday and increase fan participation. In its typical social media way, Facebook fans shared the campaign with their social circles, increasing the effectiveness of the campaign and resulting in a good uptake for the restaurant's coupons. The coupon redemption increased customer traffic to the P.F. Chang stores. This level of cross-channel activities is prevalent in online advertising. Very often, the purpose of an advertisement is to seek the customers' interest and get them to click a link, which takes them to product information, a promotion campaign, or a sign-up for some activity.

The third level optimizes the activities for an overall marketing goal. The activities may span many channels—social media sites, stores, or other ways of spreading a marketing message. Typically, a marketer is seeking a multiplier effect, in which the customers will spread the word for the marketer for a campaign. A good example comes from Starbuck's use of Facebook to decide where to introduce its pumpkin spiced latte a week before national launches in the United States and Canada. The goal in this case was to create a marketing buzz for the national campaigns. Starbucks set up a competition across its 37 million fans distributed around the world. Fans in each city participated to help their city win by carrying out activities like doing a city shout-out or a daily challenge. Chicago and Calgary were the winners in the United States and Canada, and the ultimate winner was the Starbucks marketing team.[16]

Cross-channel orchestration is a relatively new concept, and is rapidly gaining popularity due to the automation and data accessibility across channels. While the initial pilots are limited, they are showing big results in the correlational value of channels. Once properly instrumented, cross-channel marketing activities can track customer shopping across channels and offer "Next Best Action," based on the customer's current state of decision-making. For example, if the customer is conducting searches on a product, the advertising can be targeted to provide specific features of interest. Once the customer has completed the search, the campaign can turn to promotional activities. Once the customer makes a purchase, the campaign can be refocused on selling add-ons or gaining testimonials to social circles. These activities are performed today in a highly organized way in business-to-business marketing by dedicated account management teams. Technology may facilitate bringing this capability to large-scale consumer marketing.

MARKET TESTS

A formal test requires three components. First, it should be possible to find two or more nearly equivalent groups of customers to be compared. Second, a marketer should be able to conduct different campaigns in each group. Third, it should be possible to compare and contrast the results. From the observations and instrumentations described so far, we not only have the ability to design experiments but also to conduct many such experiments and optimize the marketing program based on a large number of independent experiments.

Let me take the example of Cuppa Haven described earlier in this chapter. I introduced two campaigns. The first was a postal campaign to the top 25 percent of the communities visiting a mall. The second was a poster placed at the movie theater. Both offered the same app for download, and Cuppa Haven was able to compare the results of the two campaigns to decide that the poster at the movie theater was a better campaign. In this case, the two groups of customers were the top 25 percent of the communities likely to visit a mall, and a set of moviegoers at the mall. Cuppa Haven could place a campaign targeted to each set. In both cases, the customers were able to download the app from a source provided to them. Once downloads were executed, Cuppa Haven was able to collect and compare the results.

Central to the experiment is a market test orchestration engine that can execute these tests, collect the results, and display the results to the marketer. Sophisticated market test orchestration programs run hundreds of experiments simultaneously, especially in situations in which the object of the test is an electronic product. With software configurable products, targeted campaigns, and dynamic pricing, we have all the ingredients for market tests at a large scale.

The credit-scoring industry has discovered champion-challenger testing as a way to optimize credit collection strategies. Champion-challenger is a term used to describe the way in which the existing collections strategy, known as the champion, is routinely tested against an alternative approach, known as the challenger. To ensure accuracy, the challenger should be tested in a live environment, but controlled to avoid financial loss. Thus the challenger is developed and designed using recent customer information, and is implemented on a small, statistically robust sample with the results closely monitored.

PROPOSITION

Neil McElroy was a manager at Proctor & Gamble (P&G) when he wrote his now-famous memo on May 13, 1931, to justify hiring two additional people. The 800-word memo written on his Royal typewriter created a new wave of brand management at P&G and led to McElroy's promotion to president of the company. The memo describes how a brand manager would find a trouble spot for a brand, work with the regional sales offices to fine-tune advertising, promotion, and sales functions to improve the selling of the brand, and conduct research to measure the effectiveness of these actions.[17] Over the years, P&G and other consumer marketing organizations adopted these functions for effective brand management. This was the beginning of a formalized **orchestrated marketing**.

My proposal for orchestrated marketing was also inspired by a blog written by Jerry Wind, the Lauder Professor and academic director of the Wharton Fellows Program.[18] He has also detailed his vision for network orchestration in a chapter of a book entitled *The Network Challenge: Strategy, Profit, and Risk in an Interlinked World*.[19] Today's marketing organization is made up of a set of external agents working in conjunction with a marketer. Each organization has its goals, and is working feverishly to establish business rules, which optimize these goals at silicon speed. Collaboration across these entities takes the form of revenue exchanges and contracts that determine their nature of collaboration. However, this very distributed network organization must be orchestrated by a marketing organization to get the best overall value for the marketer.

For a marketer in this distributed and federated marketplace, the big data may come from a variety of sources, both internal and external, some derived from big data created by other industries. The product ideas may get crowdsourced. The data may get organized and mined by a set of entities in a public or private cloud. Endorsement may come from a set of satisfied customers. The marketing strategies may be executed by yet another network, which may use a variety of market mechanisms to influence the customer through a collaborative process. Advertising may be executed by an auction-driven marketing place for real-time bidding of advertising space. Orchestration, as defined by Jerry Wind, has to provide three major trends—a shift in management from control to empowerment, a shift in focus from the firm to the network, and a shift in value creation from specialization to integration.[20] The orchestrator needs to be a good conductor—a great mix of dealmaker, technical wizard, and statistician.

Orchestration is provided by an organization's use of a set of technologies. There are a couple of preconditions that must be present for the orchestration to work. For example, the marketing organization must have a complete and comprehensive understanding of the customer, using data from internal and external sources. As described earlier, the marketing organization uses advanced analytics techniques to build the customer profile and collaborates with external entities as well as the customers themselves to complete the profiles. Also, customer privacy preferences must be known and incorporated into the marketing actions.

Orchestration is executed via a series of real-time workflows, which are governed by a set of policies, predictive models, and business rules. A large number of market experiments are used to test and validate the assumptions and optimize the actions based on anticipated responses from the customers and competitive forces.

In chapter 6, I will describe the technical components needed for this organization. The foundations for the technical components require a fair amount of statistical and mathematical work. They also necessitate an understanding of unstructured and qualitative analytics. In addition, the experiment design function is becoming mainstream, and is being used for hypothesis and alternative testing. The real-time execution requires adaptive components that may change with market trends.

The chief marketing officer is gaining the spotlight, and is typically the owner of the orchestration function. This is where all the marketing actions come together. Because of the technical nature of the work, marketing analytics straddles marketing and IT organizations. In chapter 7, I will discuss the organizational implications and how these changes are reshaping the marketing organizations.

Mr. McElroy, we now need a network organization to support your original memo.

6

TECHNOLOGICAL ENABLERS

INTRODUCTION

In chapter 2, I described the rapid advancement in big data analytics due to a variety of technological and environmental factors that have contributed to its acceleration. The environment has become substantially data driven, and much of the movement is due to the use of information technology (IT). A number of capabilities have grown rapidly with tremendous worldwide crowdsourced collaboration. What are these technological capabilities and how do marketers take advantage of them? How do these capabilities enable marketing analytics for large observation sets? How do they contribute to engaging customers in collaborative conversations? Which technologies enable marketers to orchestrate their efforts and focus on small micro-segments or individuals? This chapter examines these technological enablers and how they help marketers realize the propositions described in the previous three chapters.

Let me use the characteristics of big data to set the requirements first. The four V's—velocity, volume, variety, and veracity—summarize the major technological requirements. I will cast the propositions by discussing their scaling on these V's of big data. Next, I will showcase a couple of technologies in response to these V's and how they scale to big data. Last, but not least, I will discuss hybrid architectures that enable us to integrate these technologies in an end-to-end solution.

VELOCITY, VOLUME, VARIETY, AND VERACITY OF DATA

In a 2001 article, Doug Laney from the Meta group (now part of Gartner) forecast a trend in data growth and information management opportunities. He used three V's—velocity, volume, and variety—to identify the changes in information management that will give rise to big data technologies.[1] While these V's represent many characteristics, "volume" is the dominant force behind big data (see figure 6.1). A fourth V—veracity—was introduced later to represent the fact that external data could have data integrity challenges.[2] Other V's have since been added, but I will focus the discussion here on volume, velocity, variety, and veracity. It is important to understand how the information management and analytics requirements are radically altering technological choices, as the older tools are not able to scale to the data tsunami described in chapter 3. However, IT organizations invested hundreds of billions of dollars in analytics solutions in the past. It is equally important to examine how these investments can be integrated and leveraged in the new marketing analytics solutions.

Let me start with the marketer's requirements for a large **Volume** of big data. Most organizations were already struggling with increasing the size of their databases as the big data tsunami hit the data stores. According to *Fortune* magazine, we created five exabytes of digital data in record time until 2003. In 2011, the same amount of data was created in two days. By 2013, that time period was just 10 minutes.[3] While the census data volumes were hidden from most marketers as they did not have access to the individual data, the other big data sources identified in chapter 3 challenged the limited IT capabilities. More than storage, the real issue at hand is the throughput requirement. Most of the data must be carried from its source to an analytics store and then used by statistical and unstructured analytics tools. The massive fire hose required to deal with the throughput sets the IT organizations on fire.

Last year, when my basement was flooded with 80 gallons of water, I tried initially to remove the water using towels. After removing about a gallon in approximately 30 minutes, I realized it would take me more

By Gaurav Deshpande & Arvind Sathi

Figure 6.1 What is big data?

than 40 hours to clean up the rest and would cause irreparable harm to my back. When I called the professionals, they brought industrial-quality pumps to remove the water in less than one hour. IT managers are in the same predicament. The extract, transform, and load (ETL) tools designed for business intelligence using structured data were not intended for these massive volumes. IT managers can either bring a set of massively parallel pumps to pull through the throughputs or ignore the big data pools around them. Like the sponge towel I was using in my basement, data integration and storage need a new set of tooling.

A decade ago, organizations typically counted their data storage for analytics infrastructure in terabytes. They have now graduated to applications requiring storage in petabytes. This data is straining the analytics infrastructure in a number of industries. For a telco with 100 million customers, the daily location data could amount to about 50 terabytes, which, if stored for 100 days, would occupy about 5 petabytes. In my discussions with one cable company, I learned that they discard most of their network data at the end of the day because they lack the capacity to store it. However, regulators have asked most telcos and cable operators to store call detail records (CDRs) and associated usage data. For a 100-million-subscriber telco, the CDRs could easily exceed 5 billion records a day. As of 2010, AT&T had 193 trillion CDRs in its database.[4] For every CDR, there are ten registration records in their radio access networks (RAN), and for every RAN record, the deep packet inspection (DPI) data is about ten times larger. The good news is this data is like having hundreds of cameras tracking all consumer activities. The bad news is that it requires massive throughputs.

Some of this data can be aggregated or filtered at source. Let me take the example of web usage data. The data collected may include a web link, web page content, and other details that may carry nuggets of information that can be pulled out, while rest of the data can be discarded at source. However, a marketer must specify the nuggets of data that must be kept, and then a high-**velocity** engine rapidly filters the data at source to direct the meaningful data to a large data store and discards the rest. In addition, if this engine is examining the data in real time, it can also be used for providing real-time analytics, which would be useful for conversations with the customer.

There are two aspects to velocity, one representing the throughput of data and the other representing latency. Let me start with throughput, which represents the data moving in the pipes. The amount of global mobile data is growing at a 78 percent compounded growth rate, and is expected to reach 10.8 exabytes per month in 2016[5] as consumers share more pictures and videos. To analyze this data, the corporate analytics infrastructure is seeking bigger pipes and massively parallel processing. Latency is the other measure of velocity. Analytics used to be a "store and report" environment in which reporting typically contained data as of yesterday—popularly represented as "D-1." Now, the analytics is increasingly being embedded in business processes using data-in-motion with reduced latency. For example, Turn (www.turn.com) is conducting its analytics in ten milliseconds to place advertisements in online advertising platforms.[6]

These flows no longer represent structured data. Conversations, documents, and web pages are good examples of unstructured data. Some of the data, such as that coming from telecom networks is somewhat structured, but carries such a large variety of formats that it is almost unstructured. All this leads to a requirement for dealing with high **variety**. In the 1990s, as data warehouse technology was introduced, the initial push was to create meta-models to represent all the data in one standard format. The data was compiled from a variety of sources and transformed using ETL (extract, transform, load) or ELT (extract the data and load it in the warehouse, then transform it inside the warehouse). The basic premise was a narrow variety and structured content. Big data has significantly expanded our horizons, enabled by new data integration and analytics technologies. A number of call center analytics solutions are seeking analysis of call center conversations and their correlation with emails, trouble tickets, and social media blogs. The source data includes unstructured text, sound, and video in addition to structured data. A number of applications are gathering data from emails, documents, or blogs. For example, Slice provides order analytics for online orders (see www.slice.com for details). Its raw data comes from parsing emails and looking for information from a variety of organizations—airline tickets, online bookstore purchases, music download receipts, city parking tickets, or anything a consumer can purchase and pay for that hits his/her email. How do we normalize this information into a product catalogue and analyze purchases?

Unlike carefully governed internal data, most big data comes from sources outside our control, and therefore suffers from significant data integrity problems. **Veracity** represents the credibility of the data source. If an organization were to collect product information from third parties and offer it to their contact center employees to support customer queries, the data would have to be screened for source accuracy and credibility. Otherwise, the contact centers could end up recommending competitive offers that might marginalize offerings and reduce revenue opportunities. Many social media responses to campaigns could be coming from a small number of disgruntled past employees or persons employed by the competition to post negative comments. For example, we assume that "like" on a product signifies satisfied customers. But what if a third party placed the "like"?[7] Marketers as well as customers can find almost any information publicly. However, filtering for trustworthy information is an important task. Earlier, I discussed how simple rules such as the number of reviews on a Yelp page are indicators of veracity. When the marketer uses automated means to gauge market sentiments, these veracity filters are an important part of the solution to make the information trustworthy.

In the next sections, I will introduce big data technologies and discuss how they deal with large volume, velocity, variety and veracity of data.

STREAM COMPUTING TO ADDRESS VELOCITY

Traditional computing was built on a batch paradigm of observed data augmented with reported data wherever observations could not be made. Consider, for example, a call center. All the customer call information was collected at the call center and extracted, transformed, and loaded into a data warehouse, which provided trend analysis and reporting on the call center data. Some of this data was observed data. How many customers called in a particular day? What was the average wait time? What was the average handling time? Either the agent or the customer reported the rest of the data. At the end of the call, the customer had the option of reporting his/her satisfaction with the call. Did the customer provide feedback at the end of the call? How many rated the company a 1, or "very poor"? Was that trend up or down from prior days? Also, the agents provided their recollection of the call purpose and other relevant information. With any reported information, the information may not be consistently collected or properly keyed. In a call center study I carried out, I found the busy call center agent was being asked to use 87 keywords to describe a call at the end of the conversation. Most agents used the top ten keywords to codify the call. Were the calls accurately reported, or did the call center agents memorize a couple of key words and use them repeatedly because it was hard to memorize 87 keywords?

Increasingly, call centers are moving to an event-driven, continuous intelligence view of operations. This approach enables the immediate detection and correction of problems as they appear, rather than after-the-fact changes. It also allows marketers to observe and codify customer conversations using a set of tools that record observations and do not rely on the recollection of the facts by either the agent or the customer. As the conversations are carried out, relevant data can be extracted from these conversations and forwarded to the marketing organization. The process involves creating a stream-computing engine, which can observe conversations and identify relevant information during the observation.

Stream computing is a new paradigm. In "traditional" processing, one may think of running analytic queries against historical data—for instance, calculating the average time for a call last month for a call center. With stream computing, a process can be executed that is similar to a "continuous query" that keeps running totals, as observed data is collected moment by moment. In the first case, questions are asked of historical data, while in the second case, observed data is continuously evaluated via static questions. For example, stream computing can be used to observe the conversations in order to specify the mood of the caller—happy, sad, angry. It can be used to monitor customer reactions in social media to a new product launch, and all this information can be analyzed and reported—in real time!

Stream computing is best used when a marketer is dealing with a high volume or variety of data throughputs and when the data requires filters, counts, or scoring in real time. Sentiment analytics of the presidential State of the Union address is a good example of a situation in which incoming data is unstructured social media comments, and the researcher uses the incoming data to extract and report sentiment information in real time. A number of marketing activities described in chapter 4 are prime candidates for stream computing. The number of events created by the customer or the environment is staggering. For example, as a customer initiates shopping for an electronic device, marketers can observe patterns associated with him/her by analyzing a chain of events as they occur. Stream computing enables analysis and identification of the context of a customer behavior. It can examine a number of campaigns and compare them using prespecified scoring models to trigger certain actions that collaboratively influence customer actions using advertising, expert testimonial, or targeted promotions. An orchestrator can change stream-computing parameters, thereby making its action dynamic as well as adaptive to change.

There are three parallel technical concepts—complex event processing, streams, and ETL—that built the momentum and gave rise to the sophistication behind streams computing, making it the most powerful big data technology for marketing analysts. Complex event processing owes its genesis to the simulation technologies and deals with identifying complex patterns of events.[8] As marketers assemble raw data such as web searches, browse through product sites and visits to stores, they can assemble the events to recognize someone shopping for a product. Complex event processing provides the concepts for representing these events. Unix Streams[9] provided the concept of streams of data as input to a computer system. Thus, a video stream from a surveillance camera or an audio stream from a call center conversation can be ingested into a program for the detection of patterns. Last, but not least, ETL tools in business intelligence provided the mechanism for extract-transform-load of the data from a variety of operational systems.

Stream computing benefited from these technical concepts, although it is not a replacement for the respective tools that perform these functions. However, it combines the best of these concepts for the marketing problem at hand—the ability to detect patterns of events in streaming data sets and analyze the alternatives in real time. Stream computing does its activity at very high velocity, making it a tool for instant response and collaborative conversation.

Stream computing can perform three functions in real time for marketers. First, it can sense, identify, and align incoming streaming data to known customers or events, and join these events to identify the customers with a specific context and intent (for example, "customer at a store"). Second, it can categorize, count, and focus. These capabilities provide important real-time attributions to the source data, for example, "more than two web searches on a specified topic" or "a user who clicks an advertisement and goes to advertiser's site to shop." These functions use a set of dynamic parameters that are constantly updated based on deep analytics on historical data. Third, it provides capabilities to score and decide. A set of scoring models might be promoted through predictive modeling that uses historical data. These models can be scored in real time using streaming data and be used for decision-making. In addition, complex decision trees or other rule-based strategies can also be executed with run-time engines that take their rules from a business rule management system (BRMS).

Stream computing is typically performed on a massively parallel platform (MPP) to achieve high velocity and throughput. Let me discuss next the notion of an MPP system and show how these architectures support big data volumes.

ANALYTICS AND STORAGE ENGINES ON MASSIVELY PARALLEL PLATFORMS FOR HIGH-VOLUME PROCESSING

Big data usually shows up with a data tsunami that can easily overwhelm a traditional analytics platform designed to ingest, analyze, and report on typical customer and product data from structured internal sources. In order to meet the volume challenge, we must understand the size of data streams, the level of processing, and related storage issues. The entire analytics environment must have the capabilities to deal with this data tsunami and should be prepared to scale up as the data streams get bigger.

The use of massively parallel computing for tackling data scalability is showing up everywhere. In each case, the underlying principle is a distribution of workload across many processors, as well as the storage and transportation of underlying data across a set of parallel storage units and streams. In each case, the manipulation of the parallel platform requires a programming environment and an architecture, which may or may not be transparent to the applications.

Let me offer a metaphor to introduce MPP. During Thanksgiving, my son was in town, and we decided to cook together. We were in a hurry as we wrote down the grocery list. As I went to the grocery store to purchase Thanksgiving dinner items along with my wife and son, we decided to distribute the task, each person covering a couple of aisles to get part of the groceries. We decided to meet back at the checkout counter where the first one to return should get in line. In this case, by working in parallel, we cut down our overall time by a factor of three. We zigzagged through the aisles and sometime overlapped our paths, but still were far faster than one person going through the entire store with the grocery list. As we split the task, we divided the grocery list by zones, one person taking care of the vegetables, the second one taking care of frozen items, and the third dealing with wines. What I am demonstrating here is the basic principle of MPP. There is more than one agent working in parallel. We used zones to divide the task, and we combined our work at the checkout counter. Now, imagine the same work with tens, maybe hundreds, of us working together in the same fashion.

MPP is being applied to a number of areas. In the last section, I discussed stream computing. A data engineer may set a large number of processors in parallel to count, filter, or score streams of data. The parallel operation facilitates a much larger throughput. Once the data is directed to a data storage device, it may use MPP to write the data in parallel and also build its transformation using a set of parallel processes. If the data requires a sophisticated predictive data modeling activity, the statistical engine may convert the data crunching to a set of queries to be performed inside the data storage device using a set of parallel processes. For true big data performance, I may design an architecture in which each element in the data flow is in the MPP environment. Each element may choose different strategies for implementing MPP and yet provide an overall integrated MPP architecture.

Let me start with the platform for large-scale data integration. Any environment facing massive data volumes should seriously consider the advantages of MPP computing as a means to host their data integration infrastructures. MPP technologies for data integration are increasingly providing ease of setup and use, unlimited linear scalability to thousands of nodes, fully dynamic load node/pod balancing and execution, the ability to achieve automatic high availability/disaster recovery (HA/DR), and much lower price points at which comparable performance of traditional symmetric multiprocessing (SMP) shared memory server configurations can be achieved.

In stream computing implementations, continuous applications are composed of individual operators, which interconnect and operate on one or more data streams. Data streams normally come from outside the system or can be produced internally as part of an application. The operators may be used on the data to have it filtered, classified, transformed, correlated, and/or fused to make decisions using business rules. Depending on the need, the streams can be subdivided and processed by a large number of nodes, thereby reducing the latency and improving the processing volumes.

An MPP data warehouse (in the analytics engine) can also run advanced queries so that all the predictive modeling and visualization functions in the engine can be performed. The stored data is typically too large to ship to external tools for predictive modeling or visualization. The engine performs these functions based on commands that are given by predictive modeling and visualization tools. These commands are typically translated into native functions (for example, structured query language [SQL] commands), which are executed in a specialized MPP hardware environment to deal with high-volume data. Analytics engines carry typical functions for ELT (organization of ingested data using transformations), the execution of predictive models and reports, and any other data-crunching jobs (for example, geospatial analysis). The data storage architecture can be built using a two-tiered system designed to handle very large queries from multiple users. The first tier is a high-performance symmetric multiprocessing host. The host compiles queries received from business intelligence applications and generates query execution plans. It then divides a query into a sequence of subtasks, which can be executed in parallel, and it distributes the subtask to the second tier for execution. The host returns the final results to the requesting application, thus providing the programming advantages while appearing to be a traditional database server. The second tier consists of dozens to hundreds to thousands of processing units operating in parallel. Each processing unit is an intelligent query-processing and storage node, and consists of a powerful commodity processor, dedicated memory, disk drive, and field-programmable disk controller with hard-wired logic to manage data flows and process queries at the disk level.

Hadoop owes its genesis to the search engines, as Google and Yahoo required massive search capabilities across the Internet and addressed the capability of searching in parallel with data stored in a number of storage devices. Hadoop offers the Hadoop Distributed File System (HDFS) for setting up a replicated data storage environment, and MapReduce, a programming model that abstracts the problem from disk reads, and writes and then transforms it into a computation over a set of keys and values. With the open source availability, Hadoop has rapidly gained popularity.

When dealing with high volumes and velocity, we cannot leave any bottlenecks. All the processes, starting with data ingestion, data storage, and analytics and its use, must meet velocity and volume requirements. Some of these systems are designed to be massively parallel and do not require configuration or programming to enable massively parallel activities. In some cases, such as Hadoop, the parallel processing requires programming using special tools, which exploit the parallel nature of the underlying environment (in this case, HDFS). The Hadoop development environment includes Oozie, an open-source workflow/coordination service to manage data processing jobs; HBase for random, real-time read / write access to big data; Apache Pig for analyzing large data sets; Apache Lucene for search; and Jaql for query using JavaScript® Object Notation (JSON). Each component leverages Hadoop's MapReduce for parallelism; however, this elevates the skill level required for building applications. To make the environment more user-friendly, big data vendors are introducing a series of tools, such as Big Sheets from IBM, that help visualize the unstructured data.

HIGH-VARIETY DATA ANALYSIS

Blackberry faced a serious outage when its email servers were down for more than a day. I tried powering my Blackberry off and on because I was not sure whether it was my device or the CSP. It never occurred to me that the outage could be at the Blackberry server itself. When

I called the CSP, they were not aware of the problem. So I turned to one obvious source: Twitter. Sure enough, I found information about the Blackberry outage on Twitter. One of my clients told me that his vice president of customer service is always glued to Twitter looking for customer service problems. Often, someone discovers the problem on Twitter before the internal monitoring organization does. We found that a large number of junior staffers employed by marketing, customer service, and public relations search through social media for relevant information.

Traditional analytics has been focused primarily on structured data. Big data, however, is primarily unstructured, so we now have two combinations available. We can perform quantitative analysis on structured data as before. We can extract structure out of unstructured data and perform quantitative analysis on the extract quantifications. Last, but not least, there is a fair amount of nonquantitative analysis now available for unstructured data. I would like to explore a couple of techniques rapidly becoming popular with the vast amount of unstructured data and look at how these techniques are becoming mainstream with their powerful capabilities for organizing, categorizing, and analyzing big data.

Google and Yahoo rapidly became household names because of their ability to search the Web for specific topics. A typical search engine offers the ability to search documents using a set of search terms and may find a large number of candidate documents. It prioritizes the results based on preset criteria that can be influenced by how we choose the documents. If I have a large quantity of unstructured data, I can count words to find the most commonly used words. Wordle™ (www.wordle.net) supplies word clouds for the unstructured data provided to it. For example, figure 6.2 shows a word cloud for the text used in this book. The font size represents the number of times a word was used in the text.

This data can be laid out against other known dimensions. For example, IBM was working on unstructured data analytics in the Indian

Figure 6.2   A Wordle diagram of the text used in this book

market. A fairly large number of customer comments were available publicly. The IBM analysts used text analytics to study the key words being used as plotted against time. Figure 6.3 shows the results of this word count plotted against time.[10]

Once we start to categorize and count unstructured text, we can begin to extract information that can be used for qualitative analytics. Qualitative analytics can work with the available data and perform operations based on the characteristics of the data.

If we can classify the data into a set of hierarchies, we can determine whether a particular data belongs to a set or not. This would be considered a nominal analysis. If we have an established hierarchy, we can deduce the set membership for higher levels of the hierarchy. In ordinal analysis, we can compare two data items. We can deduce whether a data is better, higher, or smaller than another based on comparative algebra available to ordinal analytics. Sentiment analysis is one such comparison. For example, let us consider a statement we analyzed from a customer complaint.

"Before 12 days, I was recharged my Data Card with XXX Plan. But I am still not able to connect via internet. I have made twise complain. But all was in vain. The contact number on Contact Us page is wrong, no one is picking up. I have made call to customer care but every guy telling me…"

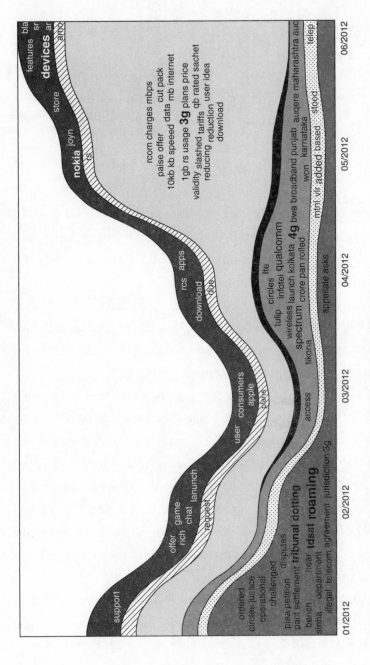

Figure 6.3 Time plot of customer blog keywords in Indian market

As humans, it is obvious to us that the sentiment of this sentence is negative. However, big data requires sentiment analysis on terabytes of data, which means we need to assign a positive or negative sentiment using a computer program. The use of words or phrases such as "I am not able," "complain," "in vain," and "no one is picking up" are examples of negative sentiments. A sentiment lexicon can be used as a library to compare words against known "positive" or "negative" sentiment. A count of the number of negative sentiments is qualitative analytics that can be performed on sentiment data, as we can differentiate between positive and negative sentiment and conclude that positive sentiment is better than negative sentiment. We can also create qualifiers such as "strong" sentiment and "weak" sentiment and compare the two sets of comments.

In typical interval-scaled data, we can assign relative values to data, but may not have a point of origin. As a result, we can compute differences and deduce that the difference between two data items is higher than another set of data items. For example, a strong positive sentiment may be better than a weak positive sentiment. However, these two data items are more similar than the pair of a strong positive and a strong negative sentiment.

PATTERN DISCOVERY

These tools take data with high variety and look for qualitative or quantitative patterns. Discovery tools include general or specialized searches across unstructured data. They carry specialized tools for machine learning for pattern recognition and can carry ontologies of different domains to make their task intelligent. With the explosion of big data, these tools advanced significantly in the qualitative analysis of unstructured data, such as social media blogs. The results of the analysis can include quantification of data, which is transferred to the analytics engine for further analysis, processing, or use. It might also result in qualitative reporting, such as semantic mapping and geospatial visualization.

EXPERIMENT DESIGN AND ADAPTIVE INTELLIGENCE

In chapter 1, I described my personal experience with watching a movie and the repeated commercials associated with the movie. In the example, I was hoping to see food processor commercials, since I was shopping for one online while watching the movie. Let me offer some details about adaptive intelligence that may provide context and user-specific variations in marketing campaign execution, leading to more focused advertising in such situations. These real-time decision engines must be based not on static rules, but on real-time analytics and are adaptive, introducing changes as they are executed. In this case, DSP should apply an adaptive bidding algorithm that changes its recommendations based on user profile, context of the video content viewing, and other contextual activities, such as my web browsing and search for food processors.

Advertisements saturate over a given number of times, after which any additional viewing is ineffective. A DMP could count the number of times an ad was displayed and decrement the likeness score for the specific ad each time it was shown, thereby favoring a different advertisement to be shown after a certain number of views. A number of sophisticated marketing experiments can be run to effectively control the saturation effect.

It seems there are two sets of input variables that constantly impact the success of advertising. The first includes search context, saturation, and response to an advertisement, and is fast moving and must be tracked in real time. The second set includes viewing habits, shopping behaviors, and other micro-segmentation-related variables, and is either static or changes gradually, accumulating over a long time period.

How would real-time adaptive analytics and decision engines acquire the background and accommodate changes to the models, while at the same time rapidly executing the engine and providing a context-dependent response? There are four components of a real-time adaptive analytics and decision engine

A sensor identifies an incoming event with a known entity. If we do not identify this entity, we can create a new one. However, if this is a known entity that we have been tracking, we will use the identifiers in the event to connect it to the previous history for this entity. The entity can be an individual, a device (e.g., a smartphone), or a known web browser identified via a previously placed cookie or pixel (see note 29 for web-tracking technologies). Under opt-in, if we placed a coupon on a smartphone and the user of the phone opted-in by accepting the coupon, we may have a fair amount of history about the individual. The analytics engine maintains a detailed customer profile based on past-identified history about the entity. The predictive modeler uses predictive analytics to create a cause-effect model, including impact of frequency (e.g., saturation in advertisement placement), offer acceptance, and micro-segmentation. The scorer component uses the models to score an entity for a prospective offer.

While sensor and scorer components may operate in real time, the analytics engine and predictive modeler do not need to operate in real time, but rather work with historical information to change the models. Returning to our example of online advertising, a cookie placed on the desktop identifies me as the movie watcher and can count the number of times an ad has been shown to me. The scorer decrements an advertisement based on past viewership for that advertisement. The analytics engine maintains my profile and identifies me as someone searching for a food processor. The predictive modeler provides a model that increases the score for an advertisement based on past web searches. The scorer picks up my context for the web search and places a food processor advertisement in the next advertisement placement opportunity. The sensor and scorer work in milliseconds, while the analytics engine and the modeler work in seconds or minutes.

Without a proper architecture, the integration of these components could be challenging. If we place all of these components in the same software, the divergent requirements for volume and velocity may choke the software. The real-time components require rapid capabilities to identify an entity, and use a number of models to score the opportunity. The task is extremely memory- and central-processing-unit (CPU) intensive and should be as lean as possible. In contrast, the analytics engine and predictive modeler may carry as much information as possible to conduct accurate modeling, including three to six months of past history and the ability to selectively decay or lower the data priority as time passes or as subsequent events confirm purchases against previously known events. I may be interested in purchasing a food processor this week and would be interested in a couple of well-placed advertisements, but the need will diminish over time as I either purchase one or lose interest.

As we engage with consumers, we have a number of methods to sense their actions, and a number of stages of engagement. A typical online engagement process may track the following stages:

1. *Anonymous customer—We* do not know anything about the customer and do not have permission to collect information.
2. *Named customer—We* have identified the customer and correlated to identification information such as device, IP address, name, Twitter handle, or phone number. At this stage, specific personal information cannot be used for individual offers because of lack of opt-in.
3. *Engaged customer—The* customer has responded to an information request or advertisement, and is beginning to shop based on offers.
4. *Opted-in customer—The* customer has given us permission to send offers or track information. At this stage, specific offers can be individualized and sent out.
5. *Buyer—The* customer has purchased merchandise or a service.
6. *Advocate—The* customer has started to "Like" the product or is posting favorably for a campaign. A real-time adaptive analytics and decision engine can help us track a customer through these stages and engage in a conversation to advance a customer from one stage to the next.

As long as the predictive modeling tool and the corresponding real-time component support the current version of predictive model markup language[11] (PMML), the predictive modelers can conduct discovery and promote the discovered model to the real-time component. Thus, as in the sports show analogy, the statistician can work in the back room and add new predictive models. The promotion process can be supported by a rigorous experiment design. The practitioners have been using the champion-challenger model for a long time in their manual promotion process. In a typical champion-challenger model, a set of models currently used in the production environment is labeled "champion." These are the current approved, accepted models for customer experience modeling. At the same time, the statisticians run an experiment design using a set of newly discovered models labeled "challengers." The experiment design is typically done using a sample small enough not to make a dent in the production environment, but large enough to be statistically significant. Let us say we randomly choose 200 households out of 10 million and use the "challenger" model for these experimental households. If the performance of the challenger is significantly better than the champion, the challenger replaces the champion for the entire population, and the process is repeated. Predictive modeling tools have used PMML to automate the champion-challenger promotion process, whereby the task of comparison, analysis, and promotion of the challenger can be performed automatically. It allows us to have better governance of the predictive models and how they are introduced or removed in the production environment.

CUSTOMER VERACITY AND IDENTITY RESOLUTION

Blogs and tweets posted by consumers on social media sites provide a wealth of information for sentiment analysis; however, this data is not structured. Consumers do not always use proper company or product names. The data contains a fair amount of slang words, and there is a mix of languages in a multiethnic, multilanguage environment. They may use a variety of words to convey positive or negative sentiments. The link to the author is not very well articulated.

We start with scant information, such as Twitter handles and unstructured references, and filter and link this data to decipher demographics, location, and other important characteristics required in making this data meaningful to a marketer.

How do we now link the Twitter handle to the customer ID? Obviously, the customer would be the best person to link them together, and customers can sometimes be incentivized to do so with product promotions or information exchange. When such direct means are not available, entity resolution technologies are providing ways to discover and resolve identities.

Identity resolution is the next step in the evolution of matching technologies for MDM. Initially developed by Jeff Jonas for the casino industry, this is a powerful technology that takes into account both normal as well as deceptive data from customers. The technology is based on a set of rules that places the probability of a match on a set of seemingly unrelated facts. As hard facts match, the probabilities are altered to reflect newly discovered information. Customer-initiated actions, such as accepting a promotion, can be hard evidence added to customer handles or user IDs, connecting them to device IDs, product IDs, or customer account information.

HYBRID SOLUTION ARCHITECTURES

The architecture components described in the previous chapter must be placed in an integrated architecture in which they can all coexist and provide overall functionality and performance consistent with our requirements. However, the requirements are at odds with each other. On the one hand, we are dealing with unstructured data discovery over very large data sets that may have very high latency. On the other hand, the adaptive analytics activities are bringing the analytics to a conversation level requiring very low latency. How do we establish an overall architecture that respects both of these components equally while establishing a formalized process for data integration? This chapter describes an integrated architecture that responds to these challenges and establishes a role for each component that is consistent with its capabilities. The architecture outlined in this chapter is the advanced analytics platform (AAP).[12]

IT organizations in all major corporations face several important architecture decisions. First, an existing infrastructure, with a large body of professionals who care for and feed the current analytics platform, is severely constrained by the growing demand for the four Vs and faces the big data tsunami. Continued investment in the current infrastructure to meet future demand is next to impossible. Second, as market forces seek new ways to create analytics-driven organizations, they are forcing massive changes in how they deal with marketing, sales, operations, and revenue management. Intelligent consumers, greenfield competitors, and smart suppliers are forcing organizations to rapidly bring advanced analytics to all major business processes. Third, the new MPP platforms, open-source technologies, and cloud-based deployment are rapidly changing how new architecture components are developed, integrated, and deployed. AAP grew under these architecture demands with four architecture principles:

1. It integrates with and caps the current analytics architecture to the mature functions, which continue to require the current warehouses and structured reporting environments. This integration includes important functions such as financial reporting, operational management, human resources, and compliance reporting. Most organizations have mature data flows, analytics solutions, and support environments. These environments will gradually change, but a radical change takes time and investment, and might not result in the biggest payback.

2. It overlays a big data architecture that shares critical reference data with the current environment and provides the necessary extensions to deal with semistructured and unstructured data. It also facilitates

complex discoveries, predictive modeling, and engines to carry the decisions driven by the insight created through advanced analytics.

3. It adds a necessary real-time streaming layer, which is adaptive, using discovery and predictive modeling components, and offers decision-making in seconds or milliseconds as needed for business execution.

4. It uses a series of interfaces to open up the data and analytics to external parties—business partners, customers, and suppliers.

I will use an analogy from sports television coverage to demonstrate how this architecture closely follows the working behavior of highly productive teams. I have always been fascinated by how a sports television production is able to cover a live event and keep us engaged as an audience using a combination of real-time and batch processing. The entire session proceeds like clockwork. It is almost like watching a movie, except that the movie is playing live with just a small time buffer to deal with catastrophic events (like wardrobe malfunctions!).

As the game progresses, the commentators use their subject knowledge to observe the game, prioritize areas of focus, and make judgments about good or bad plays. The role of the director is to align a large volume of data, synthesize the events into meaningful insight, and direct the commentators to specific focus areas. This includes replays of moves to focus on something we may have missed, statistics about the pace of the game, or details about the players. At the same time, statisticians and editors are working to discover and organize past information, some of which is structured (e.g., the number of double faults in tennis, or how much time the ball was controlled by one side in American football). However, other information that is being organized is unstructured, such as an instant replay, where the person editing the information has to make decisions about when to start, how much to replay, and where to make annotations on the screen to provide focus for the audience. The commentators have the experience and expertise to observe the replays and statistics, analyze them in real time, and explain them as they do for the game itself.

The commentators process and react to information in real time. There cannot be any major gaps in their performance. Most of the data arrives in real time, and is processed and responded to in real time as well. The director has access to all the data that the commentators are processing, as well as the commentators' responses. The director then has to script the next couple of minutes, weighing whether to replay the last great tennis shot or football catch, focus on the cheering audience, or display some statistics. In the course of these decisions, the director scans through many camera views, statistics, and replay collections, and synthesizes the next scenes of this live "movie." Behind the scenes, the statisticians and editors are working in a batch mode. They have all the history, including decades' worth of statistics and stock footage of past game coverage. They must discover and prioritize what information to bring to the director's attention.

Let me now apply this analogy to the big data analytics architecture, which consists of three analytics layers. The first is a real-time architecture for conversations; this layer closely follows the working environment of the commentators. The second is the orchestration layer that synthesizes and directs the analysis process. Last, the discovery layer uses a series of structured and unstructured tools to discover patterns and then passes them along to the orchestration layer for synthesis and modeling.

There have been four significant developments in recent years to make such platforms real-time and actionable.

Reporting versus insight: Many people believe that reports are the key mechanisms for gaining insight into data. Reporting is typically the first task for an analytics system, but it is definitely not the last. You build on reporting often by visualization of various forms that include the overlaying of geospatial visualization and the creation of new semantic models. Doing so helps you to gain insights that lead to new abstracted data. These insights can be broad, ranging from mobility patterns to micro-segments. As you gain insight, you contribute to previously unseen patterns through discovery. This pattern discovery that leads to deep insights is core to effectively using big data to transform the enterprise.

Sources of data and data integration: Merely having data does not mean you can start applying analytic tools to the data. You often must extract, transform, and load (ETL) the data before you can effectively apply these tools. Beyond ETL, it is important to integrate multiple data sources so that the analytics tools can identify key patterns. This integration is especially important given the wide variety of data sources available today. Departments create new intradepartment data everyday, including sensor, networking, and transaction data, which affect the department. The enterprise creates data such as billing, customer, and marketing data, which are essential for the enterprise to operate effectively. Third-party data also becomes critical, often sourced from social media or purchased from third-party sources. These various sources of data, which are often difficult to correlate, must be integrated to truly gain insights, which are currently not possible.

Latency and historical analytics tradeoffs: Latency that is associated with the data can often have a huge impact on how one analyzes the data and the response to the insights gained from the analytics. The perception often is that when you increase the data-gathering speed or fine-tune the hardware and software, you can move from historical analytics to real time. Historical analytics cannot often be performed in real time for a variety of reasons, including the lack of access to critical data in a synchronized manner at the right time, tools that cannot perform analytics in real time, and required dynamic model changes that are not part of existing tools for historical data. This is partly because real-time analytics introduces extra complications such as the need for logic and models to change dynamically as new insights are discovered. In addition, real-time analytics can be more expensive than historical analytics, so you must consider return on investment to justify the additional expenses.

Veracity and data governance: As mentioned earlier, Veracity represents both the credibility of the data source and the suitability of the data for the target audience. Governance deals with issues such as how to cleanse and use the data; ensure data security but still enable users to gain valuable insight from the data; and identify the source of truth when you use multiple sources for a data source and determine which is the source of truth. In most environments, data is a mixture of clean trusted data and dirty untrustworthy data. One key challenge is how to apply governance in such an environment.

SUMMARY

This chapter provided prominent technological enablers that help marketing analytics. These enablers become integrated in an advanced analytics platform to support marketing use cases such as intelligent campaigns or real-time targeted advertising. These enablers scale to the challenges posed by the large volume, velocity, variety, or veracity of data.

The enablers use a number of newly developed big data analytics techniques. To develop these solutions, marketers may need to upgrade the analytics skills in their organization. In the next chapter, I will address the skills gap and how these skills can be cultivated.

7

CHANGES TO MARKETING ECOSYSTEM AND ORGANIZATION

INTRODUCTION

We discussed three propositions in chapters 3, 4, and 5. These propositions, respectively, have major implications for marketing ecosystems and organizations. How do these propositions reshape the marketing community as we have known it in the past? Do they shrink the marketing organization or expand it? How do they reshape the departments, the agencies that marketers deal with, and the skills cultivated by these organizations over the past decades?

My attempt in this chapter is to identify three types of changes occurring in marketing organizations. First, there are major changes in the measurements and key performance indicators (KPIs). I will discuss the changes to these measures and KPIs, as well as our ability to more accurately report on the results. The second is a new wave of qualitative and quantitative skill sets needed to drive these propositions through marketing organizations. Next-generation marketers are as savvy as their customers, and are far more analytically driven. Third, the marketing business ecosystem is changing radically and I will go through the changes.

Marketing scholarship has studied the use of marketing techniques using "perfectly available" information. The aspirations all along in the marketing literature were for there to be freely available large-scale data, and the ability to take the minutest action and choreograph each action like the best New York symphony conductor. The reality was somewhere else. Marketers always craved data they did not have. The channels were at best available for broadcast, and were often controlled by gigantic contracts, and the choreography was neither feasible nor well understood. So, marketers built ways to work with the limited tools they had. The results were silos of organizations, competing budget priorities, and a fair number of skills devoted to data processing. There were miles of cubes of clerical people with calculators in the 1970s and with Excel in the 1990s, hunched over rolls of computer reports, and they massaged the data to make it ready for their decision-makers. When marketers abandoned certain markets, products, or segments, information technology (IT) could not even figure out which reports had to be retired. Reams of reports went to phantom marketing departments that no longer existed. It was hard to draw the information food chain. At one organization, I found and counted 1,500 people with "marketing analyst" titles. Most of them turned out to be pseudo-IT organizations outside the chief information officer's (CIO's) organization that were pulling data from different sources and reorganizing for their respective marketing function, each with its own set of claims on customer conversion and effectiveness, often overlapping and unverifiable.

Now, these organizations are rapidly changing. In many cases, these organizations are getting replaced with data service providers. In some cases, it is removing the walls across departments. A new breed of marketing scientist is receiving attention—the data scientist. The statistician in the back room has suddenly found the spotlight and a role in marketing strategy. Having a lot of observations radically alters the role of marketing researchers and media planners. If marketing organizations now have the ability to engage the customer in personalized micro-segmented communications, this changes the focus and role for product managers, promotion and pricing support functions, and direct marketing organizations and outsourcers.

Improved orchestration impacts product managers and advertising agencies, and brings advertising and promotion organizations closer to each other. Meanwhile, IT organizations grew in the shadow of the financial organizations and started with reporting to the chief financial officer (CFO). Over time, they graduated to the operational organizations and started reporting to the chief operating officers (COOs). The chief marketing officer (CMO) has now earned his/her right to be the next big customer or boss for the CIO, as enormous IT investments get diverted from other business groups to the CMO organization.

HOW DOES BIG DATA CHANGE MEDIA PLANNING AND MARKETING RESEARCH?

Denyse Drummond-Dunn posted a provocative blog on LinkedIn's CMO Network, "Is There Still a Need for Marketing Research Departments?" She raised a number of interesting questions. The one that I found most interesting was "Am I capable of accepting that true insight development doesn't come from one study or database, but from information integration of multiple sources?"[1] In my opinion, the biggest impact of large-scale data is directed to the marketing research organization. Not that it gets downsized, but that it has a more important role in sourcing and organizing big data.

In the past, marketers invested much more research into obtaining "reported observations" using marketing techniques. Unfortunately, a reported observation is a recollection of the facts, so it loses accuracy, and it is also dependent largely on how the question was asked. The good news is now we have a lot more "real observations," representing what actually happened. So, how do organizations get access to these real observations, and how much investment is needed before the data can be translated into marketing actions?

Take the example of media research conducted by syndicated media companies like Nielsen. Much of the media research in the past was based on small samples. Nielsen ran a panel of television viewers and another panel of shoppers, and used their data to project the results to the population using census and other secondary sources for extrapolation to the population. Today, with the availability of cable viewership data from cable set-top-boxes (STBs), cable operators are aspiring to supply their television viewership data to marketers. In the United States, their first major attempt was Project Canoe, which was formed as a consortium of six cable operators. This was in response to a Google-Echostar partnership, which used satellite viewing on the Dish network. *When it comes to advertising, Google is not shy about stating its ambitions. "We are confident we are going to revive the television advertising industry," says [Google TV's Vincent] Dureau, "by bringing new advertising to it." Already, Google is trying to make TV ads more relevant, easier to target, and cheaper to deploy. As a result, Google thinks it can attract more ad dollars from smaller businesses that may not have been advertising on TV before.*[2]

Unfortunately, the Canoe project failed in meeting its original goals. Six years later, the effort was scaled back to providing interactive advertising for video-on-demand. While Canoe was not able to get all its constituents on the same page, it did in some ways show proof of concept. It released the results of a study it jointly conducted with the American Association of Advertisers, in which a panel of 4,200 cable subscribers revealed increased product acceptance when shown interactive ads from brands like Honda, Fidelity, GlaxoSmithKline, and State Farm. According to the year-long study, 19 percent of adults 18–49 said "yes" to interactive offers, while 36 percent expressed a likelihood to purchase.[3]

The interactive video market has grown in the meantime, leading to yet another avenue for media and advertising research. Unlike linear television (which is shown on airwaves), interactive television, often termed "nonlinear," is well instrumented. Video content managers keep track of viewing details for each of their subscribers. As the nonlinear content is becoming increasingly mobile, the location of the subscriber may no longer be static. Last, but not the least, interactive viewing is more often done by an individual as opposed to a household. In my own house, interactive video has significantly reduced our traditional television viewership. The Canoe project finally found its sweet spot, and has survived and thrived as a supplier of interactive television research data.

The online viewership of data and video is closely scrutinized. Data management platforms (DMPs), such as Blue Kai, Adobe, Aggregate Knowledge, CoreAudience, Knotice, nPario, and X Plus One, track and analyze a fair amount of data from Internet viewers. Marketers are very interested in understanding a cross section of viewers, a holistic viewership data, including social, mobile, display, and search. Potentially this data can then be combined with the delivery of messages via targeted advertising or in other ways. With the rise in the mobile platform for viewership, this data can also be correlated with location data to understand and mine geographic differences.

A number of research organizations have started to collect and analyze shopping behavior. For example, I mentioned Four Square and Slice as data sources in chapters 2 and 3, respectively. A marketing research organization can use their analytics to understand shopper behavior using Four Square to map shopping, and Slice to track buying. As these organizations gather critical mass, they are able to provide valuable insight into the customer's buying process. In addition, most marketers have their own loyalty cards, which provide additional buying patterns, although focused on the brands supplied by the marketer.

Let us consider the difficult task faced by the marketing research department. By subscribing to a couple of market leaders in media research, a market researcher can easily access a vast pool of data about their customers. However, the data may have serious biases. Let me take a look at the situations I discussed earlier, in which wireless service providers and cable companies were able to offer and package mobility patterns and content usage to the retailers. If a retailer purchases a lot of smartphone, Wi-Fi, and cell tower location data from a wireless service provider to analyze traffic in a mall, the population is most likely smartphone users and not a random population. If a marketer uses audience data from a digital STB, it excludes a section of the population that is on analog STB devices. The first task for the market researcher is to identify the bias, if it is likely to skew the results. In the above examples, both samples may represent a bias toward affluent consumers. If the marketer is using these sources to model the price-sensitive buyer, the results are not going to be accurate. The second task is to use targeted research to fill in the gaps. Often, hypotheses regarding biases can easily be validated or rejected by doing directed research on the component that is underrepresented by big data. The marketer can sample portions of the observations to test the hypothesis. These examples carve out a new role for the marketing researcher—the one who integrates a variety of sources of data and adjusts the data before it is misused.

Integration of data is the next challenge. Chapter 3 showed a set of overlapping big data sources. As a researcher starts with the census data, each new data source provides additional attributes, which can be overlaid on top of census data to project a view of the population, assuming the data is properly aligned across big data sources.

Big data is also changing the measurements and KPIs in media planning and marketing research. In the good old days, advertising was focused on getting as many eyeballs ("reach") on an advertisement for a given budget and as many times as possible ("opportunities to see" or OTS) to facilitate memorization of the messaging. Since most of the programming was linear, media research organizations kept track of reach for various programs by different targets, and a media planning algorithm could maximize reach and OTS for its target segments for a given advertising budget. Click-through rate (CTR)[4] and cost-per-click (CPC) emerged as the first set of measures for online advertising. Google and others have used CPC as a measure for bidding on advertisement placement.[5] While it is the first step in the evolution of advertising effectiveness, CTR often does not result in sales. Most Internet users employ those sites for a variety of information retrievals and do not have an explicit goal or desire to click on advertisements.[6] Also, clicks are only related to browser activities, while media exposure that led to the click could be inspired elsewhere. A more convergent scenario is one in which someone may be watching television and browsing for advertisers' products on a second screen.[7] The action on one medium is dependent on exposure on the other media, and the advertising allocation must sense and correlate multiscreen viewing by customers. While there is a tendency among marketers to isolate the effectiveness of each channel in order to decide on budgets for each, the overall experience from a customer perspective is holistic.

Fortunately, the raw data supporting a gigantic set of events is gradually becoming available. The cable operators are working toward audience measurement data that can record audience viewing for specific programming and commercials. Internet viewing and related commercial exposure data is available from the data management platform suppliers. Mobility data can tell us the action in the form of physical shopping, while online shopping is available from online content analytics. In a study conducted for IAB France and SRI, the consulting firm PwC has recommended a five-level advertising effectiveness model using display, actual exposure, interaction, browsing, and engagement. PwC further specifies nine categories of indicators for measuring online performance using display, conversion, traffic, interaction, subscription, media, distribution, ROI, and posttests.[8] These measures are only associated with online advertising. A comprehensive set of measures and KPIs for multiscreen media research is still evolving.

PERSONALIZED MARKETING ACTIONS AND THE CHANGING ECOSYSTEM FOR ADVERTISING

This new market introduces a new set of organizations into the marketing ecosystem. We discussed in chapter 4 the new players in digital advertising (see figure 4.1). A DMP provides observations on viewership and past advertising; a demand-side platform (DSP) participates in auctions on behalf of the marketers or their advertising agencies; and analytics engines mine the data to create advanced audience attributions. The publishers either directly conduct auctions or use supply-side platforms (SSPs). The market forces are establishing a new ecosystem, in which DSPs protect the interests of the marketers, and SSPs protect and optimize the interest of the publishers. These optimizations are encoded in a set of algorithms, which are executed at silicon speeds to deliver the advertisements in real time (currently around 36 milliseconds).

If I know the specific audience to whom to deliver an advertisement, I can also optimize based on past purchases. As I showcased in my description of Linda's smartphone purchasing in chapter 5, a new prospect who has not purchased my products should receive a different advertisement, as compared to the customer who already purchased my product last week. To deliver these integrated capabilities, the media department must be tied to the hip with marketing research, which creates the unified customer profile. While some of the marketing spend may be directed to auctions, the rest may be done via directed contracts (for example, covering outdoor displays or traditional media), but must be optimized across the channels to deliver an orchestrated campaign.

How would the marketing organization be designed to effectively manage this changing ecosystem? The Association of National Advertisers, together with the World Federation of Advertisers and EffectiveBrands, a global marketing strategy consulting firm, has been conducting an ongoing global study—including a quantitative and qualitative survey—of senior marketing leaders over the past several months that is unprecedented in size and scope. The study, named Marketing 2020, has postulated that the winning companies will have highly integrated organizations—that is, hub-and-spoke structures, in which the CMO is in the middle, with roles akin to product manager, marketing strategies manager, advertising director, public relations (PR) manager, market-research director, and promotion director representing the spokes and the rim of the wheel. The integration and interconnectedness of this new model enable full coordination of all the constituents.[9]

MARKETING ORCHESTRATION AND THE
IMPACT ON PRODUCT MANAGEMENT

To understand changes to product management, I turned to Google. They have a massive customer base with a large number of digital connection points. Product managers have a large amount of big data at hand and an ability to test-market products at scales no one else can imagine. Ken Norton, a partner at Google Ventures, wrote a classic essay in 2005, which he claims was reinforced with his experiences at Google. Here is how he describes the personality of a product manager:

> For my part, I loved the technical challenges of engineering but despised the coding. I liked solving problems, but I hated having other people tell me what to do. I wanted to be a part of the strategic decisions, I wanted to own the product. Marketing appealed to my creativity, but I knew I'd dislike being too far away from the technology. Engineers respected me, but knew my heart was elsewhere and generally thought I was too "marketing-ish." People like me naturally gravitate to product management.[10]

Product management is an integration point between marketing, engineering, and operations. With the increased observations as described in chapter 3, as well as the orchestration capabilities as described in chapter 5, product managers have an unprecedented opportunity to design products interactively with their customers, test-market different ideas, iterate over a large number of permutations and combinations, and launch a product based on extensive feedback from the marketplace. Products can also evolve and be streamlined based on usage and feedback. All this makes the product manager a very key orchestrator across a large pool of resources.

The most innovative companies generate a greater proportion of revenue from new products and services. A study conducted by PriceWaterhouseCoopers that asked 1,700 board-level executives around the globe questions related to innovation, found that the leading innovators receive 25 percent of their revenue from innovative products and services launched in the last year compared with only 6.6 percent for the 20 percent least-innovative companies.[11] Product managers are mini-CEOs who drive innovative products through engineering, marketing, operations, and support. Consulting organizations like PwC's PRTM have been helping businesses in organizing their product management function.[12] So, how do big data, collaborative influencing, and orchestrated marketing lead to improved product management?

A fair amount of organizational division across product engineering, marketing, and operations were the result of the decision-making divisions that drove the silos across these organizations in the past. Collaboration was achieved through extensive cross-organization discussions, but resulted in long lead times. With the automation in product and customer touchpoints, product managers are now able to collect massive volumes of usage data, which can be employed to obtain rapid feedback on products' functionality, features, and components. At the same time, marketers can track shopping behavior and isolate features and components that are driving purchases. The interest can also be validated using social media data. As the smartphone purchase example illustrated in chapter 5, a marketer can identify and track shopper behavior and send specific marketing messages that influenced the purchase decision-making. In addition, the product manager can examine the product usage to investigate the extent and frequency of usage across product features. By analyzing social media, the marketer can also track positive or negative sentiments associated with each aspect. Armed with this detailed knowledge, the product manager can experiment with specific changes to the product, and test-market the changes by downloading different features on a set of beta customers. As usage patterns change, the product manager can track the changes to decipher whether they are accepted by the marketplace and whether any subtle messaging changes are needed.

This complex product management process requires an orchestrator. The product manager is closely tied to engineering, marketing, and operations, and is able to pull all the relevant data from product and customer interface to make critical product decisions. This person possesses a fair amount of respect and trust across the board, and is also able to distill key feedback from the deluge of data coming from a variety of sources, both inside and outside the corporation.

Returning to Google, statisticians embarked on a plan in 2009 to study their management. In Project Oxygen, statisticians gathered more than 10,000 observations about managers—across more than 100 variables, from various performance reviews, feedback surveys, and other reports. Then they spent time coding the comments in order to look for patterns. For much of its 13-year history, particularly the early years, Google has taken a pretty simple approach to management: leave people alone. Let the engineers do their stuff. If they become stuck, they will ask their bosses, whose deep technical expertise propelled them into management in the first place. Laszlo Bock, Google's vice president for human relations led a team to rank directives by importance. They found that technical expertise—the ability, say, to write computer code in one's sleep—ranked dead last among Google's big eight directives.[13]

DATA SCIENTISTS—WHERE DO THEY BELONG?

A marketer collects the data from all the IT trash boxes using "bit buckets" all over the organization, synthesizes it with external data, using a set of machine-learning algorithms, and then the result is a well-organized understanding of the customers, the products, and the customer interfaces. It sounds like magic! Fortunately, there is a human face to this magic, which makes sense out of all this data. It is termed the "data scientist."

As social media and big data companies went after their initial public offerings, media stories catapulted the importance of the data scientist job and the acute shortage of these workers. The data scientist grew from the unappreciated nerd in the back room to a business strategist, a quantitative genius who could consume data for lunch and dinner, and make sense of it. John Whittaker, in his blog at Dell, describes the hype about the data scientist in terms of its similarity to the webmaster in the early Internet days.

Just as there is great demand today for someone to guide companies through Big Data decisions, I recall when the No. 1 job was the almighty webmaster—the person who could ease the transitions to ecommerce and ensure the success of Internet infrastructure projects. Business leaders, in a desperate attempt to gain value that was promised by connecting their organizations to the web, paid handsomely for a webmaster with experience to get them there. Today, the same thing is occurring with the Data Scientist role. Again, a new class of technology has emerged with incredible promise and a boatload of complexities.[14]

According to a report published by McKinsey, there is a problem. "A significant constraint on realizing value from Big Data will be a shortage of talent, particularly of people with deep expertise in statistics and machine learning, and the managers and analysts who know how to operate companies by using insights from Big Data," the report said. "There will be a shortage of talent necessary for organizations to take advantage of big data. By 2018, the United States alone could face a shortage of 140,000 to 190,000 people with deep analytical skills as well as 1.5 million managers and analysts with the know-how to use the analysis of big data to make effective decisions."[15]

So, who is this mythical data scientist and how does he/she differ from the business analysts and statisticians currently employed by the marketing departments? Successful data scientists bring a triangulation of skills that are not easily blended together—an ability to understand business problems and strategy, a knack for numbers and statistical or qualitative analytics, and a dexterity in dealing with big data tools and techniques. As I watch my colleague, Tommy Eunice, who has done much of my location analytics number crunching described in chapter 3, I am amazed at how he brings all three skills to the table in a single conversation. I realized the enormity of the skill gap in my last round of recruiting for IBM. While there were a sizable number of people who claimed to be data scientists, most of them were software engineers who lacked business problem-solving skills. I have also come across many business analysts in the course of my work who understand marketing, but are too afraid to embrace quantitative techniques or new information technologies. So, how do we clone Eunice, who has all the three? Or, do we form a team with complementary skills?

To divide and conquer, we must differentiate between data scientists and data engineers. A data engineer builds the data pipes from which the data can be collected from a variety of sources and transformed so that it can be collectively analyzed. A data scientist works on the data lake and discovers insights. However, most people do a combination of these two jobs. A data engineer is typically an IT person who may report either to the IT or the marketing department, and who has strong skills in data integration. A data scientist is more than likely a consultant or a marketing department employee who has spent a fair amount of time learning machine-learning or statistics and is able to tear through massive heaps of data to find the needles in the haystacks.

The data engineer works with big data integration tools. A number of tools have been contributed to the Apache site for data sourcing, real-time data analytics, and data reorganization. In addition, big data vendors have introduced a number of proprietary tools, which work well with the open-sourced components, but offer the necessary secret sauce to realize the integrated architecture. The data engineer also works with internal and external data feeds, and has a good understanding of how these feeds can be used to identify and merge records based on selected identifiers and how the data quality can be improved. Unlike the structured business intelligence applications, especially in finance and revenue, the big data sources may be comparatively messy in their data quality. It is important for the data engineer to maintain a delicate balance between quality and latency, sometime some inaccuracy in targeting a customer may be tolerable as long it meets the latency criteria

Is the data scientist the same as the statistician? Today's big data is often unstructured and lacks the formal statistical disciplines. A data scientist must carry a fair amount of an exploratory mindset and a machine-learning background to work with unstructured data in order to glean useful structures from it. As compared to statistics, the unstructured data analytics is a relatively less understood area. The data scientist offers a blend of business and quantification skills, but may have, additionally, skills in qualitative algebra. Often, it is hard to convert unstructured data to a quantified set for statistical analysis. Techniques like graph theory deal with how masses collaborate, and it is important for the data scientist to know how to combine a variety of techniques to seek patterns from data. In my discussions with successful data scientists, I have found that most of them did not have computer science degrees. People with a functional education—engineering, business, or liberal arts, with a big dose of business experience, were more likely to be successful data scientists.

A data scientist armed with good insight can easily earn a fair amount of respect from senior management. After all, we have been data starved for decades. The data scientist brings bottoms-up insight by using real data. The knowledge of marketing function and customers is an important prerequisite in developing meaningful insight. However, marketers do require a data- and analytics-driven culture to appreciate and cultivate the data science.

"The enterprises that will achieve a competitive edge and win will have a blend of a healthy data-science culture, enterprising data scientists who can bend the ear of C-level decision makers, and the right combination of technology that will surface the data that make sense in the context of the business," says Anjul Bhambhri, vice president of development for big data projects at IBM.[16]

To help educate the community, a number of universities and big data companies are offering educational and training programs. Coursera, started by two Stanford professors offers a series of data science courses with contributions from University of Washington.[17] In addition, a number of meetups are emerging in different cities, which facilitate idea sharing among budding data scientists. As much as big data has created the demand, the community is rising to the challenge through unprecedented collaboration.

INFRASTRUCTURE, DATA, OR ANALYTICS AS A SERVICE

One way to reduce the burden of building a big data infrastructure and an army of data scientists within a corporation is to use external sources of analytics. The big data marketplace has provided a much-needed avenue for the external sourcing of analytics. A strong business model around monetization has facilitated a rapid rise in cloud-based companies offering services.

Most IT organizations would like to build the big data infrastructure within the firewalls of the organization. However, this has turned out to be a difficult task. Big data analytics requires a sizable big data infrastructure, which is hard to acquire, install, and configure. The personnel closest to the task must have appropriate data engineering and data science backgrounds. Also, there should be adequate tools for analysis. However, that means the organization procuring the tools should know which tools to acquire. The evaluation process and then the subsequent training of the individuals take time and money.

This process is not much different from any other technology that was imported in the past. However, the big data industry is rapidly evolving cloud-based services that reduce these lead times. By offering the analytics infrastructure in the cloud, the need no longer exists to evaluate, acquire, and install hardware and software. At a fairly low cost, the infrastructure can be configured in the cloud. The service provider also shows up with data scientists and data engineers who build a library of analytics and customize the library to individual needs. Often, these services can be configured in days and start producing results.

There are different types of service providers. Classic cloud providers, such as Amazon Web Service (AWS) or Softlayer, offer hardware and software assets in the cloud without any preconfigured analytics. Google analytics, Coremetrics, and Adobe are examples of analytics as a service on web traffic. A number of telcos, cable, and satellite operators are offering data as a service in which they provide data about their customers to their business partners. Cloud-based analytics companies offer social media sentiment analytics. These analytics providers bridge an important gap in breaking the ice for big data analytics. In a survey last year, I found over 30 independent programs across a large telco using a number of external, cloud-based sources. These programs provide a much-needed initial experience in using big data analytics, and as long as they are properly contained, the experience gained is invaluable.

While it is relatively easy to understand infrastructure as a service, data or analytics as a service is harder to organize. In offering data to third parties, the supplier must either aggregate the data across customers or obtain permissions to share the data. While it is easy to obtain permission against a discount, very often customers do not recall giving permission, which is often buried in pages of contract terms. Global brands are cautious in sharing data with others, as a backlash from customers can lead to irreparable damage to the brand and their mainstream business. Often, aggregation of customers is the safest approach to data sharing. Instead of providing raw data in aggregate form, the providers are also attempting analytics using the data.

Most of these are examples of analytics in which the original data was already external to a corporation. A proper analytics requires combining external data with internal data. Can we ship sensitive data to the cloud to be combined with external data by a third-party analytics as a service company? There are serious customer privacy and regulatory issues relating to data sharing across organizational boundaries. However, customer attitudes are rapidly changing with regard to the legitimate correlation of customer data, which benefits the customer. At the time of writing this book, there are two serious market experiments going on. First, the start-up activity using venture capital is experimenting with the subscription-based collection of customer data. These sources are independently collecting customer order, mobility, or other information with the premise of sharing that data at the aggregate

level with others. However, many large providers are running pilot programs to share their data either with explicit customer opt-in or blanket
opt-in, and with opt-out for those who are not interested in sharing.
Analytics as a service will evolve over time as marketers begin to share
their queries with each other and devise ways to share their data.

A NEW ROLE FOR MARKETING COMMUNICATIONS DEPARTMENTS

The Blackberry outage as described in chapter 6 shows use of social media
for discovering product problems. How would a marketing communication department use social media for customer communication?

A social media command center combines automated search and
display of consumer feedback expressed publicly on social media.
Often, the feedback is summarized in the form of "positive" or "negative"
sentiment. Once the feedback is obtained, the marketer can respond
to specific comments by entering into a conversation with the affected
consumers, whether to respond to questions about an outage or obtain
feedback about a new product offering. The marketing organization for
Gatorade, a sports drink product, decided to create a social media command center to increase consumer dialogue with Gatorade.[18] Big data
analytics can be used to monitor social media for feedback on product, price, and promotions, as well as automate the actions taken in
response to the feedback.

This may require communication with a number of internal organizations, tracking a product or service problem, and dialogue with
customers as the feedback results in product or service changes. When
consumers provide feedback, the dialogue can only be created if the
responses are provided in low latency. The automated solutions are far
better at systematically finding the information, categorizing it based
on available attributes, organizing it into a dashboard, and orchestrating a response at conversation speed. Many marketers have initiated
customer interaction with their customers using Facebook pages or
other social media activities to encourage interactive communications

with the customers. The successful ones are using social media effectively to create a buzz for the next product or the next promotion. In chapter 5, I showed some examples of situations in which these communications are the front-end for cross-channel collaboration. In a typical cross-channel collaboration, the orchestration is driven by the product managers, who use the communications department for customer interaction and coordinate their activities with the campaigns or promotions run at storefronts or other touchpoints.

EVOLUTION VERSUS REVOLUTION

The big data storm has rocked the current analytics infrastructure for many marketers. In most cases, the analytics infrastructure was not intended to deal with the volume, variety, or velocity of data anticipated from these new sources. Most marketing organizations were not equipped for handling the volumes of data, engaging in collaborative influence, and orchestrating across many organizations within and outside the corporate boundaries. The organizations may not have the right measures in place to track progress at this fine-tuned level of collaboration. The success of campaigns may be defined in silos, making it hard to collaborate across channels. The chosen tools for data integration, storage, or data mining were unable to scale to these new requirements. How does a marketing organization upgrade their current environment? The upgrade involves process, people, and technology. While it is easy for a technologist to offer a greenfield analytics environment, it may require a massive transformation of the business processes, measures, KPIs, skills, and organizational relationships. How do we deal with change at this magnitude without seriously disrupting a well-functioning organization, which may not be optimally running today, but is ill-equipped to handle the extent of change?

Earlier in this chapter, I discussed the extent of changes in the business processes, measures, collaborative objectives, external relationships, and skills. Fortunately, these changes are already being witnessed by the marketing organization today. As I studied the organizations in a number of industries, I found that marketing organizations are at different levels of maturity and that most leaders are not just at the receiving end. They are driving these changes and often leading the charge to other organizations, which support them with support processes, data, or technologies. Even where the maturity is low, marketing organizations have been able to use external services to drive significant change.

Do we start from people and process changes, or use the major shift in technology as a catalyst for organizational change? We can either start from the current marketing function and evolve into the new marketing function, making incremental changes in people, process, and technology, or make a radical departure from the past and create a new marketing analytics platform for a pilot organization and use the experiment to choreograph major changes in the organization. Both approaches have obvious pros and cons. In this section, I describe the three alternatives and discuss what would tilt us in one direction or another for a specific implementation.

With the serious investment in IT organizations, the well-organized Business Intelligence (BI) environment is the hardest to change. A typical big data analytics environment requires three significant advancements in the IT system. First, it must reduce latency by an order of magnitude, providing accessibility to data in minutes or seconds as opposed to hours or days. Second, it must increase the capacity to store data by an order of magnitude, moving from terabytes to petabytes. Third, it should have the ability to ingest external data and align it to its customers and products, and participate in external communications using the insights gathered from the analytics platform.

The big data technologies come with a significant cost advantage. The software cost is much lower because of the crowdsourced open-source components, which have also reduced the costs for commercial offerings. Because the architecture is typically built on commodity hardware and requires fewer administrators, the cost, too, is reduced by an order of magnitude. So, the good news is that we can change the IT into a self-funded model. That is, savings pay for the cost of change. However, these implementations require a commitment to big data analytics and a strong desire to migrate from the current platform. What if we have already invested a large IT budget in conventional BI? How far do we go in the first phase? Do we replace the current data warehouse architecture or augment it with big data analytics tools?

Automation is often the biggest catalyst for change. It can also be the most serious inhibitor to change. In a typical "traditional" architecture, there are a set of components for ingesting data, a set of components for storing the data, and a set of components for analyzing the data and then feeding the results into a set of actions or reports. Since all the data must be routed via a storage medium using a data warehouse, the storage, organization, and retrieval of data creates a bottleneck. Typically, the traditional approach requires a reorientation of the data from the data source to a system of record and then into a set of models conducive to analytical processing—which typically requires a number of data modelers, database administrators, and extract, transform, and load (ETL) analysts to maintain the various data models and associated keys. Changes to the business environment require changes to models, which cascade into changes across each component and require large maintenance organizations. These maintenance organizations are distributed between marketing and IT groups and must be reoriented to deal with big data. Many IT architects have already started to break away from this traditional model. Today's analytics engines do not strictly follow this paradigm, and they significantly reduce the model maintenance costs by reducing the need for representation and key-driven performance tuning.

As I study a marketing organization's plans for radical transformations, there are tea leaves available to assess the organization's maturity and will to change. Most leading organizations have adapted big data at the strategic level. How do we sense that? Here are a couple of important signs:

1. Has the organization declared data to be a strategic asset and decided on investment in data to develop a competitive marketing position? Most leaders have recognized that the "bit buckets" are full of meaningful insights, which must be stored and harvested irrespective of the cost.

2. Has the organization started to engage in a new set of business partners for the strategic alignment of marketing programs to drive the use of big data? Most leaders are using the term "monetization" as a way to define these programs. Using business partners, they are able to move rapidly toward their monetization goals, which engage customers in novel ways, as described in chapter 4.

3. Has the marketing organization developed a strong link with an IT organization? Classically, the CIOs reported to the CFO or COO. Marketing used to be a secondary objective of the CIO, while the big jobs were the revenue or Enterprise Resource Planning (ERP) implementations and operations. Very often the relationships were adversarial. By their very nature, marketers drove change in information definitions, while information technologists fought for governance and control. This is changing. In many leading organizations, marketing is the most important customer for the IT organization. The CIO may even be reporting to the CMO.

The revolutionary approach involves creating a brand-new big data analytics-driven organization. Typically, it starts with a forward-looking marketing organization that has decided to use information as a competitive strategy. The marketing organization is seeded with analytics-driven individuals and has adopted a series of KPIs to measure their performance using the power of big data.

The marketing data lake in these organizations is in the new environment, which naturally scales to the velocity and volumes of big data.

This approach has been adopted by many greenfield analytics-driven organizations. They place their large storage in the Hadoop environment and build custom analytics engines (often created using custom hardware and software) on the top of that environment to perform orchestration. The conversation layer uses the orchestration layer and integrates the results with customer-facing processes. The stored data can be analyzed using big data tools. This approach has provided stunning performance.

If an existing IT organization must be transformed to create the big data analytics environment, the cost in skill and technology transformation is substantial. It radically changes the roles and skills for the IT organization and places many more technical activities in the marketing organization. Most of the marketing organizations where this approach has worked were analytics-driven high-tech or electronic commerce companies. Analytics in these companies is not an afterthought, but a competitive edge.

In a typical evolutionary approach, big data becomes an input to the current BI platform. The data is accumulated and analyzed using structured and unstructured tools, and the results are sent to the data warehouse. Standard modeling and reporting tools now have access to social media sentiments, usage records, and other processed big data items. Typically, this approach requires sampling and processing big data to shelve the warehouse from the massive volumes. The evolutionary approach has been adopted by mature BI organizations. The architecture has a low-cost entry threshold as well as a minimal impact on the marketing and IT organizations, but it is not able to provide the significant enhancements seen by the greenfield operators. In most cases, the BI environment limits the type of analysis and the overall end-to-end velocity. All the big data flows through the new platform, while conventional sources continue to provide data to the data warehouse. We establish a couple of integration points to bring data from the warehouse into the analytics engine, which would be viewed by the data warehouse as a data mart. A sample of the new data stream data would be abstracted to the data warehouse, while most of the data would be stored using a Hadoop storage platform for discovery.

The hybrid approach provides the best of both worlds; it enables the current BI environment to function as before, while siphoning the data to the advanced analytics architecture for low-latency analytics. Depending on the transition success and the ability to evolve skills, the hybrid approach provides a valuable transition to full conversion.

SUMMARY

This chapter summarized how the marketing organization is changing to reflect the changes in the marketing function. I started with the changes to marketing research, media planning, and related metrics and key process indicators. I then discussed the changing nature of advertising and its external relationship with advertising agencies. Then, I discussed the changes to product management, and how product marketing and product engineering are coming together, driven by a need to deal with mass customization. I described the changes in skills and the increasing emphasis on data science. I discussed the new role for marketing communication in engaging and monitoring customer interactions in external media, such as social media. Finally, I discussed the changes in IT and how it can either be an inhibitor or a catalyst in forcing change.

8

CONSUMER VERSUS CORPORATE MARKETING—CONVERGENCE OR DIVERGENCE?

INTRODUCTION

Most of this book covered consumer marketing concepts and case studies. While I touched on corporate marketing, in which the customer is a corporation, this was not the main focus. Let me use this chapter to reiterate the major propositions for marketers using the context of corporate marketers, and examine the issues they face by comparing and contrasting them to consumer marketers. In using the term "corporate marketing," I will be combining three other definitions often found in the literature—industrial marketing, business marketing, and business-to-business (B2B) marketing, grouping all of them under corporate marketing.

While I have spent most of my professional life selling consumer marketing analytics to a set of marketers, my own marketing activities are those of corporate marketing. My customers are corporations, and I market and sell my products to a number of departments in these corporations—marketing, information technology (IT), finance, operations, sales, and engineering. My personal experiences has given me a good exposure to how marketing and selling work together in developing competitive positioning, understanding customer needs, building solutions, and in promoting these solutions to corporate customers of high-tech and advertising services products. Unlike in consumer marketing, I found that my work required a much closer coordination between marketing and sales, and it was often hard to divide the activities clearly between the two areas.

My second experience base is from my customers who are marketing to other corporations. Many of my customers are selling to their corporate customers. In that respect, I have been observing several variations to corporate marketing situations. First of all, there is B2B marketing, in which the customer will consume the purchased product. In particular, enterprise resource planning (ERP) systems are good examples of marketers selling ERP solutions to their customers who use ERP systems for their accounting, inventory management, or human resource departments. Second, there are business-to-business-to-consumer marketing opportunities, in which the customer is a corporation selling products to consumers. For example, AT&T sells wireless services to consumers, using wireless products from Samsung, Nokia, and Apple. The final consumer for the phones is an individual. However, Samsung and AT&T provide additional value that gets bundled with the final product. Finally, there are food chains of businesses (for example, business-to-business-to-business to consumer), in which the products become increasingly specialized by each layer of the food chain. For example, a smarter sports stadium is a final product with benefits to consumers that has a number of players in the food chain, including a stadium owner, a system integrator, a software provider, a telecom service provider, and a telecom equipment provider.

Corporate marketers are seeking and receiving a fair number of benefits from the capabilities described in chapters 3, 4, and 5. Let me offer a couple of examples to illustrate how big data has invaded the world of corporate marketing and provided marketers with a much larger number of observations, new ways of collaborative influencing, and orchestration across the marketing components. Given that corporate marketers in the past were dealing with small volumes, well-defined customer clusters, and a preintegrated sales and marketing function, they can now obtain their results much faster. While public case studies are not as prevalent, corporate markets are not lacking in their use of big data tools, and often provide useful insights to consumer marketers.

As I started to research this topic, it was hard to find published literature. This gave me the inspiration to write more from experience and prepare for a sequel, which would deal exclusively with corporate marketing based on the feedback I receive from my audience.

HOW IS CORPORATE MARKETING DIFFERENT FROM CONSUMER MARKETING?

Corporate marketing deals with marketing activities associated with corporate customers. Like consumer marketing, these marketers perform market segmentation, conduct product marketing, establish pricing, offer promotions, and build an ecosystem of endorsers and ambassadors. There are some significant deviations from consumer markets. Let me illustrate a couple of them here.

In the case of consumer markets, the decision-making is less complex and more focused on individuals. For fast-moving consumer goods, individuals make product decisions based on their own criteria and sometimes use influencers to switch brands. Some products require decision-making across a family unit, where the decision-making involves spouses or parents and children. In higher ticket items, such as houses and cars, financial institutions begin to have some influence. However, for all practical purposes, the marketing communication is between an individual or family unit and the marketer. In corporate marketing, decision-making involves one or many organizations, and could involve a large number of personnel, each representing a role, a set of decision criteria, and related experiences to evaluate their alternatives. These individuals often interact with each other over a course of time to share their observations and evaluations. Depending on the level of formalization, organizations may set up formal processes for information collection, alternative comparison, and for pooling evaluations.

Business partners, suppliers, and customers also play a role in product selection. As a result, corporate marketing invests a fair amount of time in "customer account research," which is the task of understanding customer organizations, intracorporation organizational relationships and decision-making processes, and identifying individuals who participate in the decision-making.

Corporate markets are often stratified by the size of customers. A home-based small business may require a different approach to account research than a large multinational corporation with offices in several countries. The lower end of corporate markets is not very different from high-end consumers, and is often dealt in a similar way. Let me call that "small business," which can be characterized by the number of employees, customer revenue, or any other measure associated with their business activity. The next cluster of customers may be midsize companies, often termed as small and medium enterprises (SMEs), which may have a larger business potential, but may still be too large in number for a dedicated account management function. On the high end, a marketer may be dealing with large corporations, and in some cases, signature accounts, which signify top customers with significant business potential and which enjoy a dedicated channel treatment. As a marketer stratifies the market, the higher end of the stratification may require significant attention and account-specific research, as well as specialized customer needs. Often, dedicated account teams are involved in marketing to these customers, and they piggyback on standard components from other organizations that serve mass markets or small business markets and that customize marketing strategy and programs to each large customer. At the highest level, a handful of large customers may require dedicated teams and may contribute significant revenue per customer. Each level of stratification may bring ten times the number of customers, with one-tenth the revenue per customer.

Let me describe in more detail the situation with the large customers. Large teams of sales and marketing organizations often support these large customers. Thus, it is hard to penetrate a new customer.

A significant amount of activity goes into concept marketing and account research to attract the attention of new accounts. Existing relationships shield incumbents from new entrants. A fair chunk of corporate marketing effort is in building an image for the marketer to attract new customers, using a variety of channels, including targeted mailers, trade shows, white papers, presentations, and so forth. Larger sales organizations routinely invest a fair amount of their energy into new account prospecting. These organizations fiercely compete with others to win their business. Customer account research plays a major role in focusing and sharpening the marketing campaigns during prospecting.

Customer account research is often led by a customer-facing organization. In large marketing organizations, there are a number of staff functions, who collect secondary information about the customer organizations. However, anyone who interacts with customers is the best source of formal and informal information about the customer. Executives, when hired into a new organization, routinely go to the most important suppliers to get an understanding of their new organization. The customer-facing sales and delivery personnel often encounter supporting data, which is formally and informally collected, organized, and distributed across the account team. This information is increasingly electronic, and mostly unstructured. Customer Relationship Management (CRM) vendors sell tools with which the customer account information can be converted into structured form and shared more easily. Big data analytics facilitates collection, organization, and analysis for customer account research.

Let me now move from large customers to the SME market, which have many more unconnected customers, and yet, these customers are corporations and hence involve multiple customer touchpoints and relationships that must be satisfied for a customer purchase. Unlike with a large corporation, it is probably possible to get all the decision-makers for a particular product in a conference room. The information sharing is even more interesting for the SME markets, as they deal with seemingly unconnected individuals who must find each other for collaboration in the marketplace. Because of their size, SME customers are more likely to outsource some of their organizational functions to other organizations, making those organizations part of the food chain and influencers in product purchases. Cloud IT service providers, human resource firms, and accounting firms are good examples of outsourcers who are participants and have an influential role in purchase decision-making. These customers still require significant account research, and big data sources and analytics are enablers to low-cost account research and a 360-degree view, using publicly available information,

Now let me move to customers' needs, solutions, and offerings. Each customer has its customer needs, thereby providing marketers with opportunities to define solutions and offerings. Smaller organizations are more adept at changing their requirements to assimilate and use a new offering, while large organizations are more likely to develop their custom needs and seek custom solutions in response to their needs. They also have the purchasing clout to seek major changes to an offering to meet their needs, as well as seek custom pricing to get the best deal. In most industries, I ended up finding field engineering, consulting, and integration organizations that supported product customization to standard offerings. As I was studying the manufacturing and marketing of large trucks, I found they had created a standard configuration, a custom configuration, and a goody list to deal with various customers and their feature requirements for school buses, garbage trucks, long distance trucks, and so forth. In network engineering organizations, such as Alcatel Lucent, Ericsson, and Samsung, they have large field engineering organizations co-located with their customers. Telcos often move part of their engineering functions to these organizations in order to receive custom network solutions.

Collating these field-generated requirements and creating global solutions and offerings is a challenge. In most organizations, field-developed solutions travel like "tribal knowledge" and are driven by specific people. Marketers must ingest these requirements, which are often written, using many languages, and identify common solutions that cover the needs across many countries, regions, and industries. In one such exercise, Google translator was my tool for studying a large number of artifacts, and I still struggled with understanding local cultures and buzzwords. The common process involves collating requirements across many customers and creating a common solution and related artifacts, which are then communicated to the rest of the marketplace.

Products may be configured in specific ways to deal with the lower end of the corporate market, where the products can be targeted based on a cluster of customers. For example, the health-care market for a telecom marketer may be different from retail, but a large number of hospitals may offer a cluster in SME stratification with similar needs and targeted using a health-care product line. Large corporations may, on the other hand, have specialized requirements, which can only be addressed via significant product customization. Product managers often use larger corporations to experiment with requirements, hardening the product configurations to deal with smaller corporations. Alternatively, the product may be configured for smaller customers, and a customer-facing product engineering office may work toward customizing the products for larger corporations. As customers use the products in a variety of "use cases," the sales engineering organizations support product interfaces and customizations to meet and respond to specific use cases. A medical supplier may have originally created a piece of equipment to support patients who are likely to be confined to their beds. However, as doctors and hospitals begin to use the equipment, they may find that it is beneficial for more active patients and seek product changes to meet the requirements for those patients. These changes may become included in the product once the marketing organization has studied and projected a wider demand for it. Many of the product engineering and solution development ideas discussed in chapter 4 are highly applicable to custom product development and use case marketing in corporate markets. Products are often complex and require significant marketing literature to communicate and influence corporate customers. Often, business partners collaborate to create these use cases and share them across their social marketing channels.

Pricing in corporate marketing often mirrors the product customization process. Larger corporations may require customized pricing for each engineered configuration. Often, the account teams offer specialized contracts to encourage long-term engagements and use product bundles specially created for a specific customer. Contract renewal is typically an event for special promotion offers for additional products, customization services, or discounts. Account teams are responsible for overall revenue and margin, and they wield tremendous power in discounting individual products to maximize overall long-term goals for a specific account. Product marketing organizations actively participate in developing mechanisms for custom pricing. For example, software companies often set their prices based on the number of users, number of transactions, or business outcomes. Organizationally, the usage information is studied carefully to find ways to establish pricing that can be used across a specific user group, and can be computed for each user on specific use cases. The multiyear long-term nature of business often reduces the short-term profitability tactics. Most large customers are savvy negotiators who use a network of third parties and research outfits to evaluate products, compare prices, and seek the best solution. Corporations are also multiheaded. A technical buyer may be more interested in product features, while a financial buyer may focus on pricing. The winner must optimize across various constituencies to win the customer's business. Product management and financial organizations establish discounting guidelines based on profitability criteria and use analytics to establish criteria and monitor the impacts, based on incremental gains to top- and bottom line.

A significant challenge for a marketer is to rise above the commodity pricing. Corporate customers have unique requirements that result in value-added custom changes to products. This is a perfect example of economic supply and demand. A standard product may attract a fair amount of price competition. A customized product specifically designed for a customer may be unique and attract a higher price because of its uniqueness. Extensive account research and barriers to entry protect these value-based products. A marketer must understand the trends and customizations most desired by their most lucrative targets and use customizable products to attract these customers. Increasingly, these customizations are requiring collaboration across companies. A savvy marketer has a good understanding of his/her customers' business and his/her customers' end users in order to generate novel ways of product usage. A well-designed solution customized for specific use cases commands a higher price and competitive positioning as compared to commodity products. A good marketer rises above product features to discuss solutions and use cases specific to clusters of user communities, and collaborates with these communities to develop these solutions.

In the next three sections, I will use the propositions developed in chapters 3, 4, and 5 respectively to identify how corporate marketing is changing with big data. I will use customer account research and solution marketing as two areas to examine these changes and how they impact corporate marketing.

PROPOSITION 1: BIG DATA AND ABILITY
TO OBSERVE THE POPULATION

Most consumer marketers deal with millions of customers and require ways to study and reach these customers effectively. The volume of activities is critical in such an environment. A market test for an electronic products maker may require connecting with hundreds of thousands of prospective customers, while a national launch may involve millions, and a global launch may communicate with billions. Corporate marketers deal with much smaller number of customers, but with massive variety of data. While the number of customers may be much smaller, each customer includes many decision-makers, many decision-making processes, and a variety of data sources Corporate marketers have developed ways to deal with clusters of customers, each representing a single customer, as in the case of large customers, or a homogenous group of SME customers.

Let me start with some examples of large observations for corporate marketers. Customer-facing professionals get in touch with a large number of observations about these customers. Some of these observations are very structured—revenue, billing, collections, usage, number of defects. Then, there are unstructured sources—organization charts, emails, trouble reports, Request for Proposals (RFP), Request for Information (RFI), mission statements, business plans, contracts, and so forth. Some of the information is available at client sites—contact information, organizational relationships, corporate directives shared with business partners. Social media is providing a lot more new sources of information—LinkedIn, YouTube, SlideShare, Twitter, press releases, web content, and so forth.

Another gold mine of data comes from product usage information. As this information becomes available to marketers, it can be used for mining use cases—how customers are using the products and how they differ from each other. In analyzing this data, a marketer can aggregate across employees of many customers, formulate segments, and use the segmentation information to classify customers by usage behavior. Let me give an example of a wireless telecom organization that provides service to employees of large customers. The contract may include a discounted device and service payment by the employer, and additional products and services purchased directly by the employee. Given the mobility and work-at-home provisions from most employers, many of these employees could be spending a fair amount of time working from home. What if the wireless service is not adequate at the home office? The employer may be willing to invest in a network extender product from the wireless telecom supplier, which, once added to a broadband source, can boost the wireless signal at home. The employee can use the extender to enjoy better service for all cell phones in the household, not just those that were supplied by his/her employer. The wireless telecom supplier now has a sticky customer. Even if the employee changes jobs or stops subsidizing the phone, the network extender makes it harder for him/her to switch providers. By analyzing usage and mobility information, the wireless service provider can identify employees who work from home and have poor service at home.

A corporate customer may be comprised of a large number of organizations related to each other, with the buyer-seller food chain within the corporation, and each organization in the food chain holding a different relationship with the marketer. Typically, end users, such as call center operations, sales, and so forth are supported by staff functions, and then there are supplier organizations, such as IT and procurement groups. For example, for a wireless provider like Verizon, which provides services to IBM, the end customer is a selected set of employees who use Verizon's wireless devices. The IT infrastructure that interfaces with the wireless devices is managed by the Chief Information Officer (CIO) organization at IBM, and the purchasing operations have been outsourced to a third party. The seller must deal with end users, technology providers, and the purchasing organization, as well as the procurement organization. The corporate marketer must understand the objective across these groups and devise ways to market products to each of them. For wireless services, the relationship is even more complicated, as corporations discount the device and service costs, and often employees are customers buying the upgrade and additional services. Knowledge of the food chain is absolutely the most important aspect of corporate marketing. Corporations that focus primarily on their component in the food chain rapidly become relegated to a commodity provider role, while most of the value added and extra margins are awarded to the corporation with the best end-to-end vision of the marketplace.

CRM systems traditionally provided 360 degree views using structured data. Increasingly, they are augmenting their structured customer data with social media sources, and providing analytics driven campaign tools for corporate customers.[1] Web-crawling techniques provide customer snapshots, summarizing all customer activities for a dynamic 360-degree view of an account. These views combine all significant customer activity, and may include public news, new sales, problems discovered, resolutions, and customer contact activities. This information may be available publicly or in emails, memos, IT systems, or elsewhere within the organization. Unstructured analytics tools are often used for customer research. These tools crawl through public sources, internal documents, emails, as well as structured sources to define 360 degrees view of a customer organization, covering a large number of dimensions, sources, and formulating links across these sources to facilitate account research.[2] While such views are absolutely necessary for account management, marketers can benefit from the organization of this information in one place. Without the synthesis, a marketer may invest a significant amount of time collecting customer information from a variety of sources.

With product usage and other big data sources available, corporate customers can now collect much more data about their customers. While the number of customers and contracts may be small, the number of users could be much bigger. A wireless contract with a global Fortune 50 company could easily include a large number of employees. A hotel contract for corporate travel could involve thousands of employees. Can a corporate marketer use big data analytics and borrow ideas from consumer markets for organizing segmentation information, connecting with their customer base, or engaging in marketing campaigns? Would it be possible for a corporate marketer to advertise their value-added products to a corporate employee who is using a corporate contract for basic services, but could use additional services from the marketer?

PROPOSITION 2: NEW WAYS TO INFLUENCE THE CUSTOMER

There is plenty of room for information sharing, collaborative influence, and competitive intelligence to fuel marketers' information needs. For example, YouTube is being used extensively for the public sharing of ideas, case studies, and testimonials. Many of these commercials are fairly long. For example, a popular Corning 5:32 minute long commercial has been viewed by over 23 million viewers on YouTube and has received over 17 thousand comments.[3] As the use of public sources becomes widespread, they are also turning out to be a great data mine for competitive intelligence. LinkedIn is providing a great opportunity for professionals to connect with each other. Before any meeting with customers, sales persons are often checking LinkedIn profiles of people they will be meeting. LinkedIn also provides special groups, such as "CMO Network", which facilitate group discussions among LinkedIn members on common interest topics among group members.

In the software industry, user groups have provided valuable feedback to the software marketers covering how the products are being used, as well as feedback on future directions. Often, a small number of influential users are invited to in-depth product direction discussions, in the form of "user council" or "user board". These concepts are creating non-electronic versions of exclusive clubs, where marketing ideas are shared with a selected few. The social media sites are beginning to capture this idea in the form of "velvet rope," derived from exclusive membership only gambling or dance clubs.[4] In a typical implementation, a corporate marketer uses a shared collaboration area where membership is by invitation-only. The invited members can invite others to join, thereby creating a buzz for an idea. The site provides certain privileges not available elsewhere. The membership restriction creates an exclusivity and a demand to be included. The extra privileges provide the extra value to those who join this exclusive club. The collaboration can be used for discovering new product uses, new product ideas, or for prioritizing product features and additions. The biggest value to the participants is that their use cases form the basis for new product ideas and features, thereby reducing their cost for implementing the product in their organization.

Collaboration jam is a similar concept where collaboration is facilitated in a time-boxed manner to facilitate idea sharing across

customers. In chapter 4, I covered some of the examples of collaboration jam and how these jams can be used for generating ideas. The collaboration jam facilitates a number of prospective customers to come
together. Collaboration jams can be used for generating ideas from
one customer for a solution that can be implemented across a market.
Take the example of the Covjam, a collaboration jam organized by IBM
and the city of Coventry. The three day 24 hour interactive forum was
named CovJam and generated over 2,000 posts from participants contributing their ideas and opinions. Participants debated ways Coventry
could attract inward investment, sustain employment in the local area,
personal security and how quality of life could be improved for all the
city's inhabitants.⁵ The jam benefited the city in generating new ideas,
but also helped create the vision for a "smarter city", which can be used
by other UK cities. *"Jam technology is a proven technique for drawing
on the wisdom of crowds, and capturing their enthusiasm and ideas in a
way that wouldn't be possible through traditional forms of consultation,"*
said Fraser Davidson, IBM UK Vice President for Local Government.
*"We're working to help cities not only realise their sustainability ambi-
tions but also to enable them to improve the lives of people in UK cities."*⁶

PROPOSITION 3: ORCHESTRATION FOR
CORPORATE MARKETERS

Let me continue to use the two examples—customer account research,
and solution marketing—to showcase orchestration opportunities
with corporate marketers. Customer account research is directed
towards account management. In a typical large corporation, data
about a specific customer may be shared with hundreds of corporate
marketer's employees and business partners, who collaborate to sell
a set of solutions. Account manager is the chief orchestrator. In all
the situations I have seen, the heads of account management for the
largest accounts are the most important sales personnel and often
carry Vice President, Managing Director, or General Manager titles.
They have enormous clout with all the major profit centers, and are

the ultimate decision-makers in the firm's strategy to the strategic account managed by them. They conduct their orchestration across many solutions and many sales and delivery organization to optimize their relationship with the strategic account they manage. While they receive an enormous amount of data across the corporation, they are extremely good at focusing and prioritizing marketing strategies to their account, based on the customer's best interest and toward the long-term relationship with the customer. They face the data explosion as new sources of data pour even more information onto their door-steps. They use analytics to organize and prioritize the raw data, and use collaboration tools to automate routine customer interactions. The power of analytics is in providing fast access to relevant facts associated with a situation, and in providing feedback on the quality of the routine customer interactions.

Consider the case of a wireless service provider, who provides wireless services to employees of a large corporate customer. Each employee is provided a wireless device, using a corporate contract. In many corporate situations, such contracts may be centerpiece of large contracts worth hundreds of millions or billions of dollars and require a senior-management person to manage the account management function. The account manager must be aware of service quality received by each employee, a routine customer interaction, which can be automated for data collection and collation, thereby providing the account manager with aggregate measures of performance across all employees of a corporate customer. The interaction with these employees may be in the form of emails and corporate websites, and may attract extensive blog-ging providing additional information regarding customer perceptions, problems, and accolades. Customer's IT organization is possibly collaborating with the marketer to offer new corporate apps, security and in on-boarding new employees to receive services. Periodic renewals of the contract are managed by the purchasing department, with feedback from the customer's senior management. The wireless provider must showcase continued innovation in new offerings, high service quality measures, low defect rates, and competitive pricing to keep winning new business. Big data can be used extensively for collecting metrics associated with routine interactions, and combined by the account manager to decide marketing and communication strategy. Some of the interactions will be automated. Orchestration for an innovation-driven strategic account may be very different than another interested in service quality, or a third focused on cost. Each requires a different set of data sources, activities, and foci.

Global solution marketing is an equally important marketing function. A solution marketer may study many implementations of a product to seek innovative ways of product usage, packaging, and bundling. The solution may involve joint development and marketing with business partners. Marketers may use analytics to identify solution areas, competitive activities, and case studies across many regions and to collect ideas for the solution. The solution may get discussed with business partners and customers using "velvet rope" type interactions, before general announcements. A large number of personnel may receive white papers, and presentations on solution components, with a careful selection of components based on engineering feasibility, market interest, and competitive positioning.

Consider the case of connected cars. This is a new concept jointly developed between automobile companies, telecom service providers, and software companies. A number of companies are engaged in developing the concept. Product components, including cars and wireless services, are both mature products in the marketplace. Connected cars combine these products in innovative ways to provide a new product with enormous business potential and value. A telecom provider may team with a technology company to target automotive companies in developing connected cars.[7] While the individual products involved in creating a new offering, such as a connected car, could be purchased as commodities by an integrator, the overall value is based on an integrated end-to-end solution, which commands a higher price in the marketplace. For example, AT&T is sharing its product ideas around connected cars in its Foundry and partnership with equipment manufacturers, such as Ericsson.[8] The orchestrator in this case are the solution marketing teams at wireless providers, technology providers, and automobile manufacturers. Stakes are high, as connected car is a globally applicable solution with large revenue opportunities. Marketers are orchestrating their solutions using a variety of data sources, marketing instruments, and business partners to bring their solutions to the market.

As corporate marketers have already discovered, happy customers that pay for higher valued products often prefer products that closely meet their use cases. Is it possible for consumer marketers to extend the same benefits for higher value consumer customers and offer significant customization at a higher price? With increasing product automation, such mass customizations are both feasible as well as affordable in many markets. Consumer marketers can learn from corporate marketers how these customizations can be discovered, marketed, and priced.

We seem to have a perfect opportunity for collaboration between the two sides. While consumer marketers have big volumes, their attempts at mass customization can be influenced by techniques already being practiced by corporate marketers, who are already dealing with a large variety of use cases and customers. On the other hand, as corporate marketers explore their marketing communications with their users in the corporate markets, they have something to learn from the consumer marketers.

CONCLUSIONS

Marketing analytics is radically changing, fueled by a number of market forces. The book examined three major propositions in detail that cover consumer and corporate marketing functions.

As discussed in chapter 3, marketers now have many more observations available from the entire population. These observations can be further analyzed using advanced analytics techniques to formulate insights about the context and intentions of customers. In chapter 4,

I covered how marketing programs use this insight for collaborative influence, which may be driven directly by the marketer or through a complex web of advocates and ambassadors. In chapter 5, I discussed the orchestration across marketing function and how custom marketing activities target customers based on their current status. In chapter 6, I covered a series of technological enablers for marketing analytics. In chapter 7, I covered organizational implications.

Over the past 50 years, we have witnessed a maturing of marketing function from a broad-based mass communication to targeted interaction with individuals. It is difficult to decipher whether it was the will of the marketers that improved the technology or the gift of technology that enabled the marketers to achieve these results. Irrespective of how we assign the credit, the chief marketing officer (CMO) office today has a number of unprecedented levers. The direction is towards mass customization using analytics. Those who master these levers will be the future marketing leaders.

ABOUT THE AUTHOR

Dr. Arvind Sathi is the World Wide Communication Sector architect for big data at IBM®. Dr. Sathi received his Ph.D. in business administration from Carnegie Mellon University and worked under Nobel Prize winner Dr. Herbert A. Simon. Dr. Sathi is a seasoned professional with more than 20 years of leadership in information management architecture and delivery. His primary focus has been in creating visions and roadmaps for advanced analytics at leading IBM clients in telecommunications, media and entertainment, and energy and utilities organizations worldwide. He has conducted a number of workshops on big data assessment and roadmap development.

Prior to joining IBM, Dr. Sathi was a pioneer in developing knowledge-based solutions for CRM at Carnegie Group. At BearingPoint, he led the development of enterprise integration, master data management (MDM), and operations support systems / business support systems (OSS/BSS) solutions for the communications market, and also developed horizontal solutions for communications, financial services, and public services. At IBM, Dr. Sathi has led several information management programs in MDM, data security, business intelligence, advanced analytics, big data, and related areas, and provided strategic architecture oversight to IBM's strategic accounts. He has also delivered a number of workshops and presentations at industry conferences on technical subjects, including MDM and data architecture, and he holds two patents in data masking. His first book, *Customer Experience Analytics*, was released by MC Press in October 2011, and his second book, *Big Data Analytics*, was released in October 2012. He has also been a contributing author in a number of data governance books written by Sunil Soares.

NOTES

1 INTRODUCTION

1. Matt Carmichael, *BUYographics* (New York: Palgrave Macmillan, 2013).
2. Rob Lammle, "You May Already Be a Winner! The Story of Publishers Clearing House," Mental_floss, *http://mentalfloss.com/article/30981/you -may-already-be-winner-story-publishers-clearing-house.*
3. Anupam Saxena, "TRAI Disconnects 22,769 Spammers; 161.66M Subscribers Registered With NCPR," Medianama, March 29, 2012, http:// www.medianama.com/2012/03/223-trai-disconnects-22769-spammers -161–66m-subscribers-registered-with-ncpr/.
4. Chantel Tode, "Many Marketers Unprepared as Deadline Looms for New SMS Guidelines," September 26, 2013, http://www.mobilemarketer.com /cms/news/messaging/16242.html.
5. Marketing, "IBMs CEO on Data, the Death of Segmentation and the 18-month Deadline," *Marketing Magazine,* 13 February 2013, http://www .marketingmag.com.au/news/ibms-ceo-on-data-the-death-of-segmenta tion-and-the-18-month-deadline-36359/#.Uyilyf01Ukc.
6. Elias St. Elmo Lewis, "Catch-Line and Argument," *The Book-Keeper,* February 1903, 124. Other writings by St. Elmo Lewis on advertising principles include "Side Talks about Advertising," *The Western Druggist,* February 1899, 65–66; *Financial Advertising,* (Indianapolis: Levey Bros., 1908); and "The Duty and Privilege of Advertising a Bank," *The Bankers' Magazine,* April 1909, 710–711. http://en.wikipedia.org/wiki/AIDA_(marketing).
7. I tried using the term CSP, or communications service providers, in my previous communications. Unfortunately, that is not a well-known term outside of communications companies. So, I have resorted to using the term "telecom" to describe CSPs despite their wider role and interest.
8. The video is posted on my YouTube page at https://www.youtube.com /watch?v=pPEyYPsCxZY.

2 CHANGING WINDS

1. "From Stretched to Strengthened—Insights from the Global Chief Marketing Officer Study," IBM, http://public.dhe.ibm.comRcommon/ssi /ecm/en/gbe03419usen/GBE03419USEN.PDF.

2. Richard Bishop, *Intelligent Vehicle Technologies and Trends* (Boston: Artech House, 2005), p. 300.

3. Andrew Thurlow, "Mercedes, Nokia Team Up on Smart Maps for Connected Cars," *Automotive News*, September 11, 2013, http://www .autonews.com/article/20130911/OEM06/130919958/mercedes-nokia -team-up-on-smart maps-for-connected-cars#ixzz2mvgmqsT9; Mary Branscombe, "Ford's Vision for the Connected Car," *TechRadar.Car Tech*, June 29, 2012, http://www.techradar.com/us/news/car-tech/ford-s -vision-for-the-connected-car-1086473; Kevin Fitchard, "Honda Revamps Its Link Connected Car System, Making It Very iPhone Friendly," Gigaom, December 3, 2013, http://gigaom.com/2013/12/03/honda-revamps-its -link-connected-car-system-making-it-very-iphone-friendly/.

4. "Wireless Phone Cases Dismissed," *San Francisco Call*, July 7, 1908, California Digital Newspaper Collection: http://cdnc.ucr.edu/cgi-bin/cdnc? a=d&d=SFC19080707.2.68#.

5. Martin Cooper et al., "Radio Telephone System," US Patent number 3,906,166; Filing date: October 17, 1973; Issue date: September 1975; Assignee Motorola.

6. "Multiscreen Campaign Importance Rises with Smart Device Use," eMarketer, November 25, 2013, http://www.emarketer.com/Article /Multiscreen-Campaign-Importance-Rises-With-Smart-Device-Use /1010413#FGdLFxe0Vh8h2F0O.99.

7. "Cyber Monday Goes Mobile with 55 Percent Sales Growth, Reports IBM," IBM Press Release, December 3, 2013, http://www-03.ibm.com/press/us /en/pressrelease/42661.wss#resource.

8. Statistics recorded on Appbrain's website at http://www.appbrain.com /stats/number-of-android-apps.

9. Sam Costello, "How Many Apps Are in the iPhone App Store?", About.com, http://ipod.about.com/od/iphonesoftwareterms/qt/apps-in-app-store.htm.

10. Excerpts from Yelp website: http://www.yelp.com/faq#what_is_elite_squad.

11. Geoffrey A. Fowler, "When the Most Personal Secrets Get Outed on Facebook," *Wall Street Journal*, October 13, 2012, http://online.wsj.com /news/articles/SB10000872396390444165804578008740578200224.

12. Sasha Issenberg, "How President Obama's Campaign Used Big Data to Rally Individual Voters," MIT Technology Review, December 19, 2012.) http://www.technologyreview.com/featuredstory/508836/how-obama -used-big-data-to-rally- voters-part-1/

13. Excerpts from Opensource.org website: http://opensource.org/history.

14. Excerpts from Apache website: http://httpd.apache.org/ABOUT_APACHE .html.

15. Statistics reported on the Apache website: http://en.wikipedia.org/wiki /Apache_HTTP_Server.

16. Derrik Harris, "The History of Hadoop: From 4 Nodes to the Future of Data," Gigaom, March 4, 2013, http://gigaom.com/2013/03/04/the -history-of-hadoop-from-4-nodes-to-the-future-of-data/.

17. MSLGroup, "Why Kodaveri Di Went Viral; Lessons for Marketers," Slideshare, January 13, 2012, http://www.slideshare.net/mslgroup/rhythm-correct.

18. Sony Music India, "Why This Kolaveri Di Full Song Promo Video in HD," YouTube, November 16, 2011, http://www.youtube.com/watch?v=YR12Z8f1Dh8.

19. Gautaman Bhaskaran, "Dhanush-Shruti's 3 Fails at the Box Office," HindustanTimes.com, April 18, 2012, http://www.hindustantimes.com/entertainment/regional/dhanush-shruti-s-3-fails-at-the-box-office/article1-842419.aspx.

20. William Gruger, "Psy's Gangam Style' Video Hits 1 Billion Views, Unprecedented Milestone," Billboardbiz, December 21, 2002, www.billboard.com/biz/articles/news/1483733/psys-gangam-style-video-hits-1-billion-views-unprecedented-milestone.

21. Lisa Irby, "Welcome to My Channel," YouTube, March 18, 2013, http://www.youtube.com/watch?v=JfH5nuioMe4

22. "Internet Ad Revenues At $20.1 Billion Hit Historic High for Half-Year 2013, Up 18% over Same Time in 2012, According to IAB," IAB, October 9, 2013, http://www.iab.net/about_the_iab/recent_press_releases/press_release_archive/press_release/pr-100913.

23. Ki Mae Heussner, "The Backstory on the Most Frequently Cited Chart in Digital Media," Gigaom, June 5, 2012, http://gigaom.com/2012/06/05/the-backstory-on-the-most-frequently-cited-chart-in-digital-media/.

24. Amir Efrati, "Online Ads: Where 1,240 Companies Fit In," *Wall Street Journal,* June 6, 2011, http://blogs.wsj.com/digits/2011/06/06/online-ads-where-1240-companies-fit-in/.

25. Miller et al., "Sentiment Flow through Hyperlink Networks," Stanford University, http://www.stanford.edu/~cgpotts/papers/sentiflow.pdf#!.

26. Now part of IBM's Enterprise Marketing Management group. For details, see http://www-01.ibm.com/software/marketing-solutions/coremetrics/.

27. See details of Radian6's offering at their website: http://www.salesforcemarketingcloud.com/products/social-media-listening/.

28. Offers a command center. For details, see http://www.attensity.com/products/attensity-command-center/.

29. "Adobe Chases Elusive Profit in $67.3 Billion Cloud Market," Information Management, December 9, 2013, http://www.information-management.com/news/adobe-chaese-proft-in-cloud-10025131–1.html.

30. Anick Jesdanun, "FTC Finalizes Privacy Settlement with Facebook," *USA Today,* August 10, 2012, http://www.usatoday.com/tech/news/story/2012-08-10/ftc-facebook-privacy/56934670/1.

31. Paul Elias, "NSA Surveillance Documents Released by Officials Show Misuse of Domestic Spying Program," *Huffington Post,* September 9, 2013, http://www.huffingtonpost.com/2013/09/10/nsa-surveillance-documents_n_3902208.html.

32. As stated on Apple web site and updated as of March 1, 2014. See https://www.apple.com/legal/privacy/en-ww/ for details.

33. Phil Nickinson, "Sprint says it's no longer collecting analytics via Carrier IQ," Android Central, December 16, 2011, http://www.androidcentral.com/sprint-says-its-no-longer-collecting-analytics-carrier-iq

34. AT&T privacy policy, stated on AT&T website at http://www.att.com/gen /privacy-policy?pid=13692.

35. Drew Fitzgerald, "Yahoo Passwords Stolen in Latest Data Breach," *Wall Street Journal*, July 12, 2012, http://online.wsj.com/news/articles/SB10001 424052702304373804577522613740363638.

36. Adam Tanner, "The Web cookie is dying. Here is the creepier technology that comes next," *Forbes Magazine*, June 17, 2013, http://www.forbes.com /sites/adamtanner/2013/06/17/the-web-cookie-is-dying-heres-the-creepier -technology-that-comes-next/

3   FROM SAMPLE TO POPULATION

1. Kenneth Cukier and Viktor Mayer-Schonberger, *Big Data: A Revolution That Will Transform How We Live, Work, and Think* (Boston, New York: Eamon Dolan/Houghton Mifflin Harcourt, 2013).

2. Mike Mancini, "The US Census, why counting counts for business," Nielsen Newswire, May 18, 2010, http://www.nielsen.com/us/en/newswire/2010 /the-u-s-census-why-counting-counts-for-business.html.

3. John King, "Romney campaign plans fail to launch," CNN video, November 6, 2012 http://www.cnn.com/video/data/2.0/video/bestoftv /2012/11/07/2012-elections-king-romney-chances.cnn.html.

4. Karen Narasaki et al., "State of Asian American consumer," Nielsen report, Quarter 3, 2012, http://www.nielsen.com/content/dam/corporate/us /en/microsites/publicaffairs/StateoftheAsianAmericanConsumerReport .pdf.

5. Jonathan Taplin, "Social Sentiment: Are You Listening?", YouTube, August 28, 2012, http://www.youtube.com/watch?v=YGlhCCwYVh4.

6. Jonathan Taplin, "Social Sentiment Analysis Changes the Game for Hollywood," Building a Smarter Planet, November 30, 2012, http://asmarter planet.com/blog/2012/11/social-sentiment-analysis-could-change-the -game-for-hollywood.html.

7. The results are displayed on the bing website at http://www.bing.com /politics/stateoftheunion.

8. Adam Ostrow, "Inside Gatorade's Social Media Command Center," Mashable, June 15, 2010, http://mashable.com/2010/06/15/gatorade-social -media-mission-control/.

9. Noam Cohen, "It's Tracking Your Every Move and You May Not Even Know," *New York Times*, March 26, 2011, http://www.nytimes.com/2011/03/26 /business/media/26privacy.html?_r=0.

10. See geohash.org for more details.

11. This table is derived by computing the impact of each byte in geohash on the distance it represents. The logic is explained in http://en.wikipedia .org/wiki/Geohash.

12. Sambit Sahu, Arvind Sathi, Thomas Eunice, Mathews Thomas, Ken Kralick, "A Deep Dive into 'First of a Kind' Big Data Telco Solution," IBM Information on Demand Conference, Las Vegas, November 3–7, 2013, http://www.slideshare.net/arvindsathi/location-analytics-applications -and-architecture.

13. Chris Loza et al., "Explore the Advanced Analytics Platform, Part 3: Analyze Unstructured Text Using Patterns," IBM Developer Works, http://www.ibm .com/developerworks/library/ba-adv-analytics-platform3/index.html.

14. "No Fault Found," a white paper available on CiQ's website at http://www .carrieriq.com/documents/no-fault-found-whitepaper/7386.

15. The product description for Huawei's F256 device for broadband Internet connectivity is available at the Huawei website at http://www.huaweidevi ceusa.com/products/features?id=59.

16. Arvind Sathi, *Customer Experience Analytics: The Key to Real-time, Adaptive Customer Relationships* (Boise: MC Press, 2012).

17. Joel S Dubow, "Point of View: Recall Revisited: Recall Redux," *Journal of Advertising Research*, May / June 1994, G&R Research and Consulting, http:// www.gandrllc.com/reprints/pointofviewrecallrevisitedrecallredux.pdf.

18. Hayley Tsukayama, "Apple iPhone 5s, iPhone 5c Set New Sales Record," *Washington Post*, September 23, 2013. http://www.washingtonpost.com /business/technology/apple-iphone-5s-iphone-5c-set-new-sales-record /2013/09/23/e232699a-244c-11e3-ad0d-b7c8d2a594b9_story.html.

4   FROM BROADCAST TO COLLABORATION

1. Martha Rogers and Don Peppers, *The One to One Future* (Kindle Locations 141–142). Crown Business. Kindle Edition, 2000–01–07.

2. Tom Vanderbilt, "The Science behind the Netflix Algorithms That Decide What You'll Watch Next," *Wired*, August 7, 2013, http://www.wired.com /underwire/2013/08/qq_netflix-algorithm/.

3. "The Netflix Prize Rules," posted on Netflix website at http://www.netflix prize.com/rules.

4. "Netflix Prize Update," posted on Netflix website, http://blog.netflix .com/2010/03/this-is-neil-hunt-chief-product-officer.html

5. Tom Vanderbilt, "The Science behind the Netflix Algorithms That Decide What You'll Watch Next," *Wired*, August 7, 2013, http://www.wired.com /underwire/2013/08/qq_netflix-algorithm/.

6. Lance Whitney, "Internet Ad Sales Hit a New High of $9.6 Billion," www .cnet.com, June 3, 2013. http://news.cnet.com/8301-1023_3-57587353-93 /internet-ad-sales-reach-new-high-of-$9.6-billion/.

7. "US Stays atop Global Ad Market, But Others Rank Higher Per Capita," *eMarketer Newsletter*, September 26, 2013. http://www.emarketer.com /Article/US-Stays-Atop-Global-Ad-Market-Others-Rank-Higher-per-Capita /1010248.

8. Kuang-Chih Lee, Burkay Orten, Ali Dasdan, and Wentong Li, "Estimating Conversion Rate in Display Advertising from Past Performance Data." *www.turn.com*. http://www.turn.com/sites/default/files/kdd2012.pdf.

9. Quynh Cline, "The Turn Digital Audience Report," Turn, Inc. July–September 2013, http://e.turn.com/GDAOct2013.

10. Mark Walsh, "Mobile Showrooming Leads to In-Store Sales," Mediapost News, February 26, 2013, http://www.mediapost.com/publications/article /194372/mobile-showrooming-leads-to-in-store-sales.html#ixzz 2M7Qti5wQ.

11. Pamela Robertson, "Embracing the Cross-channel Marketing Imperative," Experien Marketing Services, June 10, 2013, http://www.experian.com /blogs/marketing- forward/2013/06/10/embracing-the-cross-channel -marketing-imperative/.

12. Based on discussions with Ashish Bisaria, Senior Vice President, Customer Experience, Manheim.

13. Abigail R. Esman, "China's $13 Billion Art Fraud—And What It Means For You," *Forbes*, August 13, 2012, http://www.forbes.com/sites/abigailes man/2012/08/13/chinas-13-billion-art-fraud-and-what-it-means-for -you/.

14. Brad Stone, "Why Redfin, Zillow, and Trulia Haven't Killed Off Real Estate Brokers." *Bloomberg BusinessWeek*, March 7, 2013, http://www.business week.com/articles/2013–03–07/why-redfin-zillow-and-trulia-havent -killed-off-real-estate-brokers#p1.

15. From iStockphoto website, FAQs, http://www.istockphoto.com/faq/how -to-use#faq-how-works.

16. Jeff Howe, "The Rise of Crowdsourcing," *Wired*, June 2006, http://www .wired.com/wired/archive/14.06/crowds.html.

17. A Typepad blog by Jeff Howe, "Crowdsourcing: A Definition," June 2, 2006, http://crowdsourcing.typepad.com/cs/2006/06/crowdsourcing_a .html

18. For details, see www.armycocreate.com.

19. The term Collaborative Innovation™ is an IBM trademark.

20. "Welcome to the IBM Jam events page," https://www.collaborationjam.com.

21. "Habitat Jam Knowedge Gallery," http://www.globaldialoguecenter.com /exhibits/backbone/index.shtml.

22. Dan Lyon, "Must See: This Google Reunion Video Is Bringing People to Tears," Hubspot, November 14, 2013, http://blog.hubspot.com/uattr/must -see-this-google-reunion-video-is-bringing-people- to-tears#!

23. Google India, "Things You Can Do on Your Mobile," posted on YouTube, September 1, 2010, http://www.youtube.com/watch?v=puTOLyDwOxY.

24. Rob Van Den Dam, "Global Telecom Consumer Survey," IBM Institute for Business Value, 2011, http://www.slideshare.net/belive/telecommunications -in-2015-its-a-brand-new-game-rob-van-den-dam-20-oct-2011.

25. Brandon Gutman, "Fischer-Price on Connecting with Moms in the Digital World," *Forbes*, September 13, 2012. http://www.forbes.com/sites/market share/2012/09/13/fisher-price-on-connecting-with-moms-in-the-digital -world/.

5   FROM SILO'ED TO ORCHESTRATED MARKETING

1. Infographics, "Breaking down Google's 2011 revenues," Wordstream, 2011, http://www.wordstream.com/articles/google-earnings.

2. PricewaterhouseCoopers, IAB Internet Advertising Revenue Report" IAB, October 9, 2013, http://www.iab.net/research/industry_data_and _landscape/adrevenuereport.

3. Sasha Issenberg, "How President Obama's Campaign Used Big Data to Rally Individual Voters," *MIT Technology Review*, December 19, 2012,

http://www.technologyreview.com/featuredstory/509026/how-obamas
-team-used-big-data-to-rally-voters/.

4. Adam Zorne's profile information is available on MDM Community website at http://mdmcommunity.ning.com/profile/AaronZornesTheMDM Institute. A number of his reports are available at the MDM Institute website.

5. US Securities and Exchange Commission web site provides summary information as well as links to relevant details on the Sarbane Oxley act of 2002. http://www.sec.gov/about/laws.shtml#sox2002.

6. David Waddington, "Styles and Architectures for Master Data Management," Information Management, July 10, 2009, http://www.information-manage ment.com/issues/2007_60/mdm_styles-10015711–1.html.

7. John Brodkin, "Casino Insider Tells (Almost) All about Security," Network World, March 7, 2008, http://www.networkworld.com/news/2008/030708 -vegas-insider.html; Daniel Weisberg, "Jeff Jonas: Chief Scientist Interview," Online behavior, http://online-behavior.com/emetrics/jeff-jonas-ibm.

8. Travis Korte, "5 Q's with Jeff Jonas, Chief Scientist of IBM Entity Analytics Group," Center for Data Innovation, December 8, 2013, http://www .datainnovation.org/2013/12/5-qs-with-jeff-jonas-chief-scientist-of-ibm -entity-analytics-group/.

9. Jacqui Cheng, "Netflix Settles Privacy Lawsuit, Ditches $1 Million Contest," Arstechnica, March 10, 2010, http://arstechnica.com/tech-policy/2010/03 /netflix-ditches-1-million-contest-in-wake-of-privacy-suit/.

10. Garland Grammer, Shallin Joshi, William Kroeschel, Arvind Sathi, Sudir Kumar, and Mahesh Viswanathan, "Obfuscating Sensitive Data While Preserving Data Usability," USPTO Patent Number 20090132419. United States Patent and Trademark Office: http://www.uspto.gov.

11. William Kroeschel, Arvind Sathi, and Mahesh Viswanathan, "Masking Related Sensitive Data in Groups," USPTO Patent Number 20090132575. United States Patent and Trademark Office: http://www.uspto.gov.

12. Based on conversations with Philip Smolin, Senior Vice President, Turn, Inc.

13. Scenario is based on a showcase developed by IBM's Global Solution Center.

14. PF Chang, "A Facebook coupon offer drives foot traffic into PF Chang's," Facebook Success Stories, 2012, http://www.facebook-successstories.com /pf-changs/.

15. For details about Clickfox, visit their website at http://www.clickfox.com.

16. Starbucks, "Starbucks engages fans by letting them choose which city gets the pumpkin spice latte first, Facebook Success Stories, 2012, http://www .facebook-successstories.com/starbucks/.

17. Sean Duffy, "The McElroy "Brand Man" memo turns 80," Brand Rants, May 11, 2011, http://www.brandrants.com/brandrants/2011/05/mcelroy -brand-man-memo.html.

18. Jerry Wind, "Orchestration as the new managerial model in the digital age," Google Think Insights, April 2012, http://www.google.com/think /columns/orchestration-as-the-new-managerial-model.html.

19. Jerry Wind, Victor Fung, and William Fung, "Network Orchestration: Creating and Managing Global Supply Chains Without Owning Them," ch. 17 in *The Network Challenge: Strategy, Profit, and Risk in an Interlinked World*, Paul R. Kleindorfer and Jerry Wind (eds.) (New York: Prentice Hall, 2009).
20. Ibid.

6   TECHNOLOGICAL ENABLERS

1. Doug Laney, "3D Data Management, Controlling Data Volume, Velocity, and Variety," Meta Group File: 949, February 6, 2001, http://blogs.gartner.com/doug-laney/files/2012/01/ad949-3D-Data-Management-Controlling-Data-Volume-Velocity-and-Variety.pdf.
2. For details, see Villanova University's website and their definition of big data, as stated on their home page titled "What Is Big Data?", http://www.villanovau.com/university-online-programs/what-is-big-data/.
3. "What Data Says About Us," *Fortune,* September 24, 2012, p. 163.
4. "Top 10 Largest Databases in the World," March 17, 2010. http://www.comparebusinessproducts.com/fyi/10-largest-databases-in-the-world.
5. "Statshot: How Mobile Data Traffic Will Grow by 2016," August 23, 2012, http://gigaom.com/mobile/global-mobile-data-forecast.
6. Kate Maddox, "Turn Ad Inspired by 'Mad Men,'" www.btobonline.com, July 16, 2012.
7. Ben Grubb, "Can't Buy Love Online? 'Likes' for Sale," www.stuff.co.nz, August 24, 2012.
8. David Luckham, *The Power of Events: An Introduction to Complex Event Processing in Distributed Enterprise Systems* (City of publication: Addison Wesley, 2013).
9. James Cox, *The Magic Garden Explained: The Internals of UNIX System V Release 4, an Open-systems Design* (New York: Prentice-Hall, 1994).
10. Arvind Sathi, *Big Data Analytics* (Boise: MC Press, 2012).
11. Alex Guazelli et al., "PMML: An Open Standard for Sharing Models," *The R Journal*, 1 (1), May 2009, http://journal.r-project.org/archive/2009-1/RJournal_209-1_Guazzelli+et+al.pdf.
12. Arvind Sathi, et al, "Explore the advanced analytics platform," IBM Developer Works, September 24, 2013, http://www.ibm.com/developerworks/library/ba-adv-analytics-platform1/.

7   CHANGES TO MARKETING ECOSYSTEM AND ORGANIZATION

1. Denyse Drummond-Dunn, "Does your Organisation Really Need a Market Research Department? And in the Future?", c³ Centricity, Oct 11, 2013, http://www.c3centricity.com/blog/future-of-maret-research-department.
2. Erick Schonfeld, "Project Canoe: Cable companies paddle to catch up to Google in targeted TV ads," Tech Crunch, March 10, 2008, http://techcrunch.com/2008/03/10/project-canoe-cable-companies-paddle-to-catch-up-to-google-in-targeted-tv-ads/.

3. Daniel Frankel, "Why Canoe abandoned ship with interactive TV ads," Gigacom, Feb 22, 2012, http://paidcontent.org/2012/02/23/419-why-canoe-abandoned-ship-with-interactive-tv-ads/.

4. Google, "Clickthrough rate (CTR), Google Support, March 16, 2014, https://support.google.com/adwords/answer/2615875?hl=en.

5. Google, Actual cost per click (CPC), Google Support, March 16, 2014, https://support.google.com/adwords/answer/6297?hl=en.

6. Conrad Feldman, "Display ad clickers are not your customers," Slideshare, March 22, 2013, http://www.slideshare.net/hardnoyz/display-ad-clickers-are-not-your-customers.

7. IBM Big Data & Analytics, "Helping marketers and advertisers adapt to ever changing audience measurement," Youtube, Jan 21, 2013, http://www.youtube.com/watch?v=QwxAvU2_BSk&feature=youtu.be.

8. PriceWaterhouseCoopers, "Measuring the effectiveness of online advertising," PWC website, March 16, 2014, http://www.pwc.com/en_GX/gx/entertainmentmedia/pdf/IAB_SRI_Online_Advertising_Effectiveness_v3.pdf.

9. Jennifer Rooney, "Here is what the marketing organization of the future should look like," Forbes, Oct 4, 2013, http://www.forbes.com/sites/jenniferrooney/2013/10/04/heres-what-the-marketing-organization-of-the-future-should-look-like/.

10. Ken Norton, "How to hire a product manager," Kenneth Norton's personal website, March 16, 2014, https://www.kennethnorton.com/essays/productmanager.html.

11. PriceWaterhouseCoopers, "Breakthrough Innovation and Growth," March 16, 2014, www.pwc.com/innovationsurvey.

12. Michael E McGrath, *Next Generation Product Development: How to Increase Productivity, Cut Costs, and Reduce Cycle Times* (New York: McGraw-Hill, 2004).

13. Adam Bryant, "Google's Quest to Build a Better Boss," *New York Times*, March 12, 2011, http://nyti.ms/pkqWuf.

14. John Whittaker, "Is Today's Data Scientist 1995's Webmaster?", Direct2Dell, http://fw.to/UkDWVbe

15. http://www.mckinsey.com/insights/business_technology/big_data_the_next_frontier_for_innovation.

16. Dan Woods, "IBM's Anjul Bhambhri on What Is a Data Scientist," *Forbes*, February 16, 2012, http://www.forbes.com/sites/danwoods/2012/02/16/ibms-anjul-bhambhri-on-what-is-a-data-scientist/.

17. University of Washington, "Introduction to data science," on the Coursera website, https://www.coursera.org/course/datasci.

18. Valerie Bauerlein, "Gatorade's 'Mission': Sell More Drinks," *Wall Street Journal*, September 13, 2010, http://online.wsj.com/news/articles/SB10001424052748703466704575489673244784924; Adam Ostrow, "Inside Gatorade's Social Media Command Center," Mashable Social Media, June 15, 2010, http://online.wsj.com/news/articles/SB100014240527487034667004575489673244784924. See also the YouTube video at http://www.youtube.com/watch?v=InrOvEE2v38.

8 CONSUMER VERSUS CORPORATE MARKETING— CONVERGENCE OR DIVERGENCE?

1. Setlogik, "360 degree customer view in salesforce.com," YouTube, February 4, 2012, http://www.youtube.com/watch?v=Lr5XLP1G2PQ.
2. IBM Big Data & Analytics, "Big data use case #2: enhanced 360 degree view of the customer," YouTube, April 9, 2013, http://www.youtube.com /watch?v=yk2uk1_Oiuw.
3. Corning Incorporated, "A day made of glass, made possible by corning," YouTube, February 7, 2011, http://www.youtube.com/watch?v=6Cf7IL _eZ38.
4. John Zahn "Use Social Media to create a buzz: The Velvet Rope," Omnibeat, February 20, 2012, http://www.omnibeat.com/the-velvet-rope/.
5. John Galvez, "Coventry City Council and IBM Successfully Demonstrate New Approach to Public Engagement," IBM Press Release, July 27, 2010, http://www-03.ibm.com/press/uk/en/pressrelease/32208.wss.
6. Ibid.
7. Alex Konrad, "IBM and Sprint team up on smarter connected cars that send data when engine's off," Forbes Magazine, July 10, 2013, http://www.forbes .com/sites/alexkonrad/2013/07/10/ibm-sprint-connected-cars-partnership/.
8. AT&T, "The AT&T Drive Studio for Connected Car Solutions and Technology," YouTube, January 6, 2014, https://www.youtube.com/watch? v=49onIYrk5GU.

INDEX